Letts

GCSE SUCCESS

VISUAL REVISION GUIDE

CHEMISTRY FOUNDATION

Author

Emma Poole

CONTENTS

EARTH MATERIALS

METALS

STRUCTURE AND BONDING

CHEMICAL CHANGE

LIMESTONE

Limestone is a <u>sedimentary</u> rock. It is mainly <u>calcium</u> <u>carbonate</u>. It can be <u>quarried</u> and cut into blocks which can be used for <u>building</u>.

If the limestone is powdered it can be used to <u>neutralise</u> the <u>acidity</u> in <u>lakes</u> <u>caused</u> <u>by</u> <u>acid</u> <u>rain</u> and to <u>neutralise</u> <u>acidic</u> <u>soils</u>.

HEATING LIMESTONE

- When <u>limestone</u> (<u>calcium</u> <u>carbonate</u>) is <u>heated</u> <u>it</u> <u>breaks</u> <u>down</u> to form <u>quicklime</u> (<u>calcium</u> <u>oxide</u>) and <u>carbon</u> <u>dioxide</u> (CO_2).

calcium carbonate $\Rightarrow$ calcium oxide + carbon dioxide
$CaCO_3(s)$ $\Rightarrow$ $CaO(s)$ + $CO_2(g)$

This is an example of a thermal decomposition reaction.

- The <u>quicklime</u> (<u>calcium</u> <u>oxide</u>) reacts with water to form <u>slaked</u> <u>lime</u> (calcium hydroxide). A solution of slaked lime is known as <u>limewater</u>.

calcium oxide + water $\Rightarrow$ calcium hydroxide
$CaO(S)$ + $H_2O(l)$ $\Rightarrow$ $Ca(OH)_2(s)$

- Because they are both bases, slaked lime can be used in the same way as powdered limestone in lakes and on soils, but it works much more quickly.

Examiner's Top Tip
Learn the equation involved when limestone is heated.

REACTION SUMMARY

limestone (calcium carbonate) → *heat* → quicklime (calcium oxide) → *add water* → slaked lime (calcium hydroxide)

↓ carbon dioxide gas

OTHER USES OF LIMESTONE

GLASS
Glass can be made by <u>heating</u> a mixture of <u>limestone</u> (calcium carbonate), <u>sand</u> (silicon dioxide) and <u>soda</u> (sodium carbonate) until the mixture melts.

CEMENT
Cement is produced by <u>roasting</u> <u>powdered</u> <u>clay</u> with <u>powdered</u> <u>limestone</u> in a rotating kiln.

If the <u>cement</u> is mixed with <u>water</u>, <u>sand</u> <u>and</u> <u>rock</u> <u>chippings</u> a slow chemical reaction produces the <u>rock-like</u> <u>concrete</u>.

Concrete is hard and cheap and is widely used in building.

Examiner's Top Tip
Remember – calcium oxide and calcium hydroxide are both bases, so they can neutralise acidic lakes and soils.

QUICK TEST

1. What is the main chemical in limestone?

2. What type of rock is limestone?

3. What is powdered limestone used for?

4. What is formed when limestone is heated?

5. What is the equation for this reaction?

6. What type of reaction is this?

7. What is formed when quicklime (calcium oxide) is reacted with water?

8. What is the equation for this reaction?

9. How is glass made?

10. How is cement made, and what can it be made into?

1. Calcium carbonate
2. Sedimentary
3. To neutralise acidity in soils/lakes.
4. Quicklime (calcium oxide) and carbon dioxide
5. $CaCO_3(s) \longrightarrow CaO(s) + CO_2(g)$
6. Thermal decomposition
7. Slaked lime (calcium hydroxide)
8. $CaO(s) + H_2O(l) \longrightarrow Ca(OH)_2(s)$
9. By heating limestone, sand and soda
10. by roasting powdered clay and limestone; it can be made into concrete.

IGNEOUS ROCKS

- All **igneous** **rocks** are formed from **molten** **rock** which has **cooled** and **solidified**. Molten rock below the surface of the Earth is called **magma**, above the Earth's surface it is called **lava**.

- **Igneous** **rocks** are very hard and have **crystals**.

- **Extrusive** **igneous** **rocks** have **small** **crystals** because they have formed very quickly above ground. Basalt is an example of an extrusive igneous rock.

granite has large crystals

- **Intrusive** **igneous** **rocks** have **large** **crystals** because they solidified slowly below the ground. Granite is an example of an intrusive igneous rock.

Examiner's Top Tip
There are quite a lot of facts here – make a table with these headings: Rock Type, Characteristics, Formation and Examples, and include all the information in the table.

SEDIMENTARY ROCKS

- *Sedimentary rocks tend to be crumbly and sometimes contain fossils. Sandstone and limestone are examples of sedimentary rocks.*

sandstone

limestone

- *Sedimentary rocks form from layers of sediment found in seas or lakes. Over millions of years these layers are buried by further sediment. The weight of these layers squeezes out the water and the particles become cemented together.*

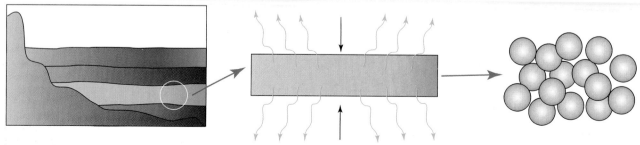

layers of sediment build up

water is squeezed out

the particles of rock become cemented together

METAMORPHIC ROCKS

Metamorphic <u>rocks</u> are usually <u>hard</u> and may contain <u>banded</u> crystals. Metamorphic rocks are formed by <u>high temperatures</u> and <u>pressure</u> on existing rocks. Metamorphic rocks can be formed when rock is <u>stressed</u> as mountains are formed or when <u>hot magma</u> comes into contact with rock, causing alteration of the existing rocks.

schist (schist and gneiss are examples of metamorphic rocks)

ROCKS CAN BE CLASSIFIED INTO 3 GROUPS:
IGNEOUS, SEDIMENTARY AND METAMORPHIC ROCKS

QUICK TEST

1. Which type of rock is formed when molten rock cools and solidifies?

2. Which type of rock is the hardest?

3. Which sort of igneous rock has small crystals and was formed quickly?

4. Which sort of igneous rock has large crystals and was formed slowly?

5. Give two examples of igneous rocks.

6. Which type of rock may contain fossils?

7. Over what sort of time period do sedimentary rocks form?

8. Give two examples of sedimentary rocks.

9. Which two factors can cause existing rock to be changed into metamorphic rock?

10. Give two examples of metamorphic rocks.

10. Schist, gneiss
9. Heat, pressure
8. Sandstone, limestone
7. Millions of years
6. Sedimentary
5. Basalt and granite
4. Intrusive
3. Extrusive
2. Igneous
1. Igneous

CLUES IN ROCKS

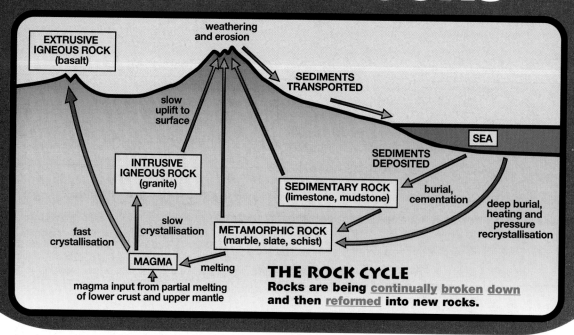

weathering and erosion

EXTRUSIVE IGNEOUS ROCK (basalt)

SEDIMENTS TRANSPORTED

slow uplift to surface

SEA

SEDIMENTS DEPOSITED

INTRUSIVE IGNEOUS ROCK (granite)

SEDIMENTARY ROCK (limestone, mudstone)

burial, cementation

deep burial, heating and pressure recrystallisation

fast crystallisation

slow crystallisation

METAMORPHIC ROCK (marble, slate, schist)

MAGMA   melting

magma input from partial melting of lower crust and upper mantle

THE ROCK CYCLE
Rocks are being <u>continually</u> <u>broken</u> <u>down</u> and then <u>reformed</u> into new rocks.

SEDIMENTARY ROCKS

- Sedimentary rocks are formed when particles are <u>deposited</u> by <u>water</u>, <u>wind</u> or <u>ice</u>.
- The <u>rocks</u> build up in <u>layers</u>.
- The <u>youngest</u> rocks are usually found on <u>top</u> <u>of</u> <u>older</u> <u>rocks</u>.

youngest rock

oldest rock

IGNEOUS ROCKS

<u>Igneous</u> <u>rocks</u> can also give useful information about the <u>geological</u> <u>history</u> <u>of</u> <u>an</u> <u>area</u>. Igneous rocks sometimes intrude into <u>existing</u> <u>sedimentary</u> <u>rocks</u>.

This can be used to <u>date</u> <u>the</u> <u>age</u> <u>of</u> <u>the</u> <u>rocks</u>. The igneous rock must be younger than the sedimentary rock that it cuts across.

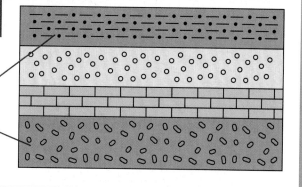

igneous intrusion (younger than the sedimentary rocks)

older sedimentary rocks

RIPPLE MARKS

- <u>**Ripple**</u> <u>**marks**</u> (parallel ridges) are sometimes seen in sedimentary rocks. They are <u>**formed**</u> <u>**by**</u> <u>**the**</u> <u>**movement**</u> of currents and waves as the particles were <u>**deposited**</u> <u>**in**</u> <u>**the**</u> <u>**sea**</u>.

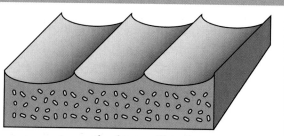

ripple marking

FOLDING

Rock layers may become bent due to the forces they are subjected to. This is called faulting:

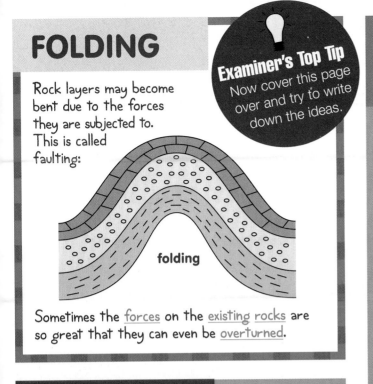
folding

Sometimes the <u>forces</u> on the <u>existing</u> <u>rocks</u> are so great that they can even be <u>overturned</u>.

FAULTING

Sedimentary rocks can be subjected to <u>great</u> <u>forces</u> as the <u>plates</u> <u>move</u>. The rock may be subjected to so much force that it <u>snaps</u>. This is called faulting:

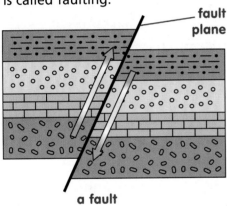
fault plane

a fault

DISCONTINUOUS DEPOSITION

Sedimentary rocks can give a lot of information about what was <u>happening</u> <u>at</u> <u>the</u> <u>time</u> <u>the</u> <u>rock</u> <u>was</u> <u>formed</u>.

A <u>discontinuous</u> <u>deposition</u> shows that there was a <u>period</u> <u>of</u> <u>erosion</u> followed by a later period where sediment was again <u>deposited</u>.

These discontinuities can occur over large areas and indicate that there was a <u>relative</u> <u>movement</u> <u>of</u> <u>the</u> <u>land</u> <u>and</u> <u>the</u> <u>sea</u>.

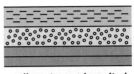
sediments are deposited

sediments are folded

time

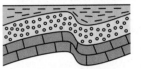
sea level drops, deposition of sediment ceases and erosion occurs

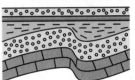
sea level rises and new sediments are deposited

QUICK TEST

1. How are sedimentary rocks turned into metamorphic rocks?
2. What type of rock is formed when magma is crystallised quickly?
3. What type of rock is formed when magma is crystallised slowly?
4. Why are younger rocks normally found on top of older rocks?
5. What causes ripple marks on rocks?
6. An igneous intrusion is found to have cut across sedimentary rock, what can be deduced about the ages of the rocks?
7. Why are some rocks faulted or folded?
8. Sketch what happens when a sedimentary rock is faulted.
9. Sketch what happens when a sedimentary rock is folded.
10. Sketch the stages which lead to discontinuous deposition.

10. Refer to diagram above
9. Refer to diagram above
8. Refer to diagram above
7. Great forces acting on the rocks may cause them to bend or snap.
6. The igneous rock is younger than the sedimentary rock.
5. The movement of currents/waves.
4. They are deposited later, on top of existing rock.
3. Intrusive igneous rocks
2. Extrusive igneous rocks
1. Through heat and pressure

FOSSIL FUELS

Fossil fuels are <u>coal</u>, <u>oil</u> and <u>natural</u> gas.

FORMATION OF COAL, OIL AND NATURAL GAS

Fossil fuels are formed over <u>millions of years</u>. They are the fossilised remains of <u>dead plants and animals</u>.

<u>Plants and animals</u> died and fell to the sea or swamp floor.

The remains were quickly covered by <u>sediment</u>.

In the absence of <u>oxygen</u> the remains did not <u>decay</u>.

As the layers of sediment <u>increased</u> the remains became <u>heated and pressurised</u> (squashed).

After millions of years <u>coal</u>, <u>oil</u> and <u>natural gas</u> are formed.

Dead plants falling into swamps form <u>coal</u>, while tiny dead sea creatures and plants form <u>oil and natural gas</u>.

Burning fossil fuels is an exothermic process – giving out a lot of heat.

Fossil fuels are <u>non-renewable</u>. They take millions of years to form, but they are being used up very quickly.

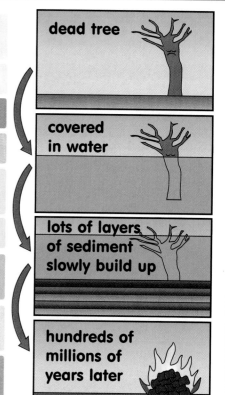

dead tree

covered in water

lots of layers of sediment slowly build up

hundreds of millions of years later

HOW SCIENTISTS HELP FIND OIL

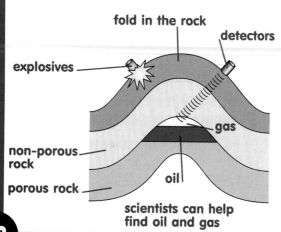

fold in the rock

detectors

explosives

gas

non-porous rock

oil

porous rock

scientists can help find oil and gas

Oil is a scarce and valuable resource. Scientists are involved in locating new reserves. The first clue that oil may be present is a fold in the rocks. If the area looks as if it may contain oil, scientists will carry out a seismic survey. This is used to analyse the structure of the rocks below ground. Small explosions at the surface send out shock waves. Each layer of rock reflects the shock waves back to the detectors. By analysing the results the scientists can find if the area could still contain oil. However, they never know for sure that oil is there until they drill through the rock. So it is a risky and expensive business.

FRACTIONAL DISTILLATION OF CRUDE OIL

- Crude oil is a <u>mixture of</u> <u>hydrocarbons</u>.
- Hydrocarbons are molecules that contain only <u>hydrogen</u> <u>and</u> <u>carbon</u> <u>atoms</u>.
- Some of the hydrocarbons have <u>very</u> <u>short</u> <u>chains</u> of carbon atoms.
- These hydrocarbons are <u>runny</u>, <u>easy</u> <u>to</u> <u>ignite</u> and have <u>low</u> <u>boiling</u> <u>points</u>.
- Other hydrocarbons have much longer chains of carbon atoms. These hydrocarbons are more <u>viscous</u> (less runny), <u>harder</u> <u>to</u> <u>ignite</u> and have <u>higher</u> <u>boiling</u> <u>points</u>.
- This means that long chain hydrocarbons are <u>not</u> <u>useful</u> <u>as</u> <u>fuels</u>.

Fractional distillation can be used to split mixtures of <u>hydro-carbons</u> (<u>crude</u> <u>oil</u>). In the fractionating column the bottom is kept very hot, while the top of the column is much cooler. The <u>smallest</u> <u>molecules</u> (those with the lowest boiling points) boil off first and <u>rise</u> <u>to</u> <u>the</u> <u>top</u> <u>of</u> <u>the</u> <u>column</u>. The other fractions are collected at different points down the column.

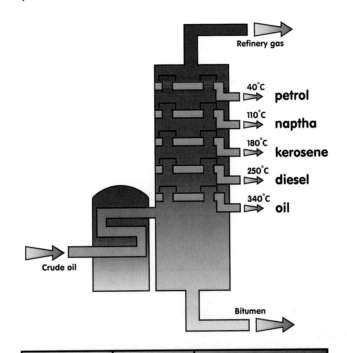

No. of carbon atoms in hydrogen chain	Temperature	Fraction collected
3	Less than 40°C	Refinery gas
8	40°C	Petrol
10	110°C	Naptha
15	180°C	Kerosene
20	250°C	Diesel
35	340°C	Oil
50+	Above 340°C	Bitumen

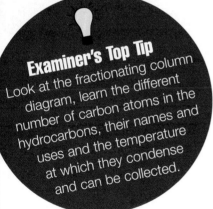

Examiner's Top Tip
Look at the fractionating column diagram, learn the different number of carbon atoms in the hydrocarbons, their names and uses and the temperature at which they condense and can be collected.

Examiner's Top Tip
Remember – plants and animals are only fossilised if no air reaches them; if oxygen does reach them the remains will decay.

QUICK TEST

1. Over what kind of time scale are fossil fuels formed?
2. Why are they named fossil fuels?
3. Why do some sediments not decay?
4. From what is coal formed?
5. From what are oil and natural gas formed?
6. Which elements do hydrocarbons contain?

6. Carbon and hydrogen
5. Tiny sea creatures and plants
4. Dead plants
3. Because no oxygen is present.
2. They are the fossilised remains of plants/animals.
1. Millions of years

ALKANES

All hydrocarbons have a **spine** of **carbon atoms**.

Molecules that belong to the **alkane family** contain no carbon–carbon double covalent bonds.

They are **saturated** hydrocarbons.

Name	methane	ethane	propane	butane
chemical formula	CH_4	C_2H_6	C_3H_8	C_4H_{10}
structure	H H-C-H H	H H H-C-C-H H H	H H H H-C-C-C-H H H H	H H H H H-C-C-C-C-H H H H H

ALKENES

The alkene family of hydrocarbons do contain carbon–carbon double bonds.

They are unsaturated hydrocarbons.

name	ethene	propene
chemical formula	C_2H_4	C_3H_6

structure	H H
	   C = C
	H H
		H H
		  C = C – C–H
		H H H

CRACKING

- The <u>large</u> <u>hydrocarbons</u> separated during the <u>fractional</u> <u>distillation</u> of crude oil are <u>not</u> very useful.
- <u>Cracking</u> can break down these large hydrocarbons into <u>smaller</u>, more <u>useful</u> molecules.
- There is <u>more</u> <u>demand</u> and therefore a <u>higher</u> <u>price</u> for the smaller hydrocarbons like <u>petrol</u> compared with larger molecules like <u>lubricating</u> <u>oil</u>.

CRACKING

Crude oil contains a <u>mixture</u> of <u>hydrocarbons</u>.

INDUSTRIAL CRACKING

Cracking is an example of a thermal decomposition reaction. Large, less useful molecules are broken down into smaller, more useful ones using heat and a hot aluminium oxide catalyst.

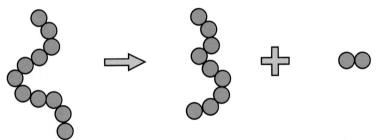

long hydrocarbons are not very useful as fuels

a useful fuel

ethene – used to make plastics

The ethene is useful because it contains a double bond between two carbon atoms. This means that it is very reactive. Ethene can be used to make plastics and other useful substances.

QUICK TEST

1. How are large hydrocarbons obtained from crude oil?
2. Why are the long hydrocarbons cracked to form smaller ones?
3. How are large hydrocarbons cracked industrially?
4. What is the catalyst that is used?
5. What is the ethene produced by cracking used for?

5. Plastics
4. Hot aluminium oxide
3. They are heated over a catalyst
2. Smaller hydrocarbons are more useful.
1. Through fractional distillation.

PLASTICS

LOTS OF SMALL HYDROCARBONS CAN BE JOINED TOGETHER TO MAKE ONE BIG MOLECULE.

MAKING PLASTICS

Hydrocarbons can be separated by fractional distillation.
The cracking of large hydrocarbons can be used to obtain small reactive molecules (as well as other molecules which can be useful as fuels).
If these small reactive hydrocarbon molecules are heated under pressure with a catalyst, they can be joined together to form a large plastic molecule.
The small hydrocarbons are called monomers.
The large molecules formed when lots of monomers are joined together are called polymers ('poly' meaning lots).
If the monomer is ethene, the resulting polymer is called poly(ethene) (meaning lots of ethenes).

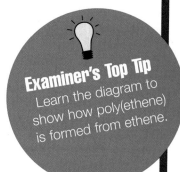

Examiner's Top Tip
Learn the diagram to show how poly(ethene) is formed from ethene.

Similarly, many propene monomers can be joined together to make poly(propene).

USES OF PLASTICS

POLY(ETHENE)

- Poly(ethene) is **cheap** and **strong**.
- It is used for plastic **bags** and **bottles**.

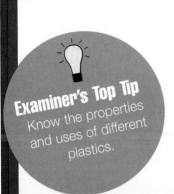

POLY(CHLOROETHENE), PVC

- PVC is **rigid** and is used for building materials such as **drain pipes**.
- With plasticisers added, it is used for **wellingtons** and **mackintoshes**.

POLY(PROPENE)

- Poly(propene) is **strong** and has a high **elasticity**.
- It is used for **crates** and **ropes**.

POLY(STYRENE)

- Poly(styrene) is **cheap** and can be **moulded** into different shapes.
- It is used for **packaging** and for plastic **casings**.

QUICK TEST

1. Name the monomer used to make poly(ethene).
2. Name the polymer formed when many propene molecules are joined together.
3. Name two properties of of poly(ethene)?
4. Give two uses of polyethene.
5. What are the properties of poly(propene)?
6. What can it used for?
7. What are the properties of PVC?
8. What can it used for?
9. What are the properties of poly(styrene)?
10. What can it be used for?

1. Ethene
2. Poly(propene)
3. It is cheap and strong.
4. Bags and bottles
5. It is strong and has a high elasticity.
6. Crates and ropes
7. It is rigid.
8. Drain pipes, wellingtons and mackintoshes
9. It is cheap and easily moulded.
10. Packaging and plastic casing

EVOLUTION OF THE ATMOSPHERE

FORMATION OF THE ATMOSPHERE

THE FIRST BILLION YEARS
- During the first billion years of the Earth's life there was enormous <u>volcanic</u> <u>activity</u>.
- Volcanoes belched out <u>carbon</u> <u>dioxide</u> (CO_2), <u>steam</u>, <u>ammonia</u> (NH_3) and <u>methane</u> (CH_4).
- The atmosphere was mainly <u>carbon</u> <u>dioxide</u> and there was very little <u>oxygen</u> (like the modern-day atmospheres of Mars and Venus).
- The water vapour <u>condensed</u> to form the early oceans.

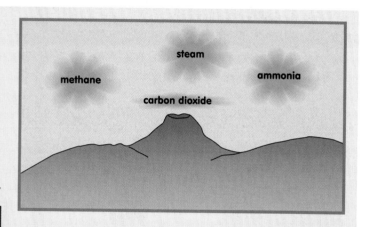

LATER
- During the next <u>two</u> <u>million</u> <u>years</u> plants evolved and began to cover most of the Earth.
- The plants grew well in the carbon-dioxide rich atmosphere and steadily <u>removed</u> carbon dioxide and produced <u>oxygen</u> (O_2).

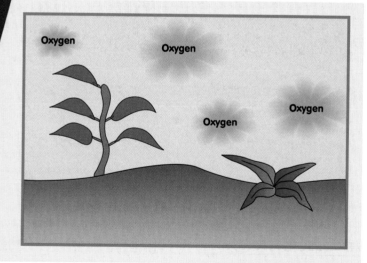

- Most of the carbon from the carbon dioxide gradually became locked up as <u>carbonates</u> and <u>fossil</u> <u>fuels</u> in sedimentary rocks.
- The ammonia in the early atmosphere reacted with oxygen, releasing <u>nitrogen</u>.

Examiner's Top Tip
Practise covering up these pages and then try to write the details down.

The composition of the Earth's atmosphere today is:
- about 80% nitrogen
- about 20% oxygen
- small amounts of other gases such as carbon dioxide, water vapour and noble gases, e.g. argon.

This has not always been the case. Throughout the history of the Earth the composition of its atmosphere has changed and evolved.

20%

80%

HOW ARE WE AFFECTING THE ATMOSPHERE TODAY?

- The burning of fossil fuels can affect the atmosphere.

- Most fuels contain both <u>carbon</u> and <u>hydrogen</u> and many also contain a little <u>sulphur</u>.

- When <u>carbon</u> is <u>burnt</u>, <u>carbon</u> <u>dioxide</u> (CO_2) is released into the atmosphere.

- When <u>hydrogen</u> <u>is</u> <u>burnt</u>, <u>water</u> <u>vapour</u> (H_2O) is released (water is an <u>oxide</u> of hydrogen).

- When <u>sulphur</u> <u>is</u> <u>burnt</u>, <u>sulphur</u> <u>dioxide</u> (SO_2) is formed.

QUICK TEST

1. Roughly how much of the atmosphere is made up of oxygen?

2. What is the main gas in the atmosphere today?

3. What other gases are present in small amounts in today's atmosphere?

4. Which gases formed the Earth's early atmosphere?

5. Which was the main gas present?

6. How did the evolution of plants affect the Earth's atmosphere?

7. What happened to most of the carbon dioxide in the early atmosphere?

8. Name the three elements that can be found in fuels.

9. Name the compound formed when hydrogen is burnt.

10. Name the compound formed when carbon is burnt in plenty of oxygen.

1. 20%
2. Nitrogen
3. Carbon dioxide, water vapour, noble gases
4. Carbon dioxide, steam, ammonia, methane
5. Carbon dioxide
6. Removed carbon dioxide, produced oxygen
7. Became locked up in sedimentary rocks
8. Carbon, hydrogen and sulphur
9. Water (vapour)
10. Carbon dioxide

ACID RAIN

- Fossil fuels may contain some **sulphur**.
- When these fuels are burnt **sulphur dioxide** is produced and released into the atmosphere.
- This gas dissolves in rainwater to produce **acid rain**.
- Acid rain can affect the environment, damaging statues and buildings as well as trees, animals and plants.

CFCs AND THE OZONE LAYER

- Chlorine is used in the production of CFCs, (chlorofluoro-carbons).
- CFCs were widely used in aerosols, fridges and as solvents.
- When these gases escape into the atmosphere they can trigger a chain reaction which breaks up ozone.
- Ozone is a molecule made of three oxygen atoms and is important because it filters out harmful ultraviolet rays. These rays can cause skin cancer and damage crops.
- The ozone is broken down to form oxygen molecules, O_2.
- CFCs have now been widely replaced in products and their production is now very limited.

Examiner's Top Tip
The ozone layer problems are caused by CFCs.

CARBON MONOXIDE

Carbon monoxide can also cause problems.
- When fossil fuels are burnt in insufficient oxygen, carbon monoxide (CO), is produced instead of carbon dioxide (CO_2).
- Carbon monoxide is colourless, odourless and very poisonous.
- Faulty gas appliances can produce this gas, which is dangerous and claims many lives each year.

CARBON DIOXIDE AND THE GREENHOUSE EFFECT

- The <u>greenhouse effect</u> is slowly heating up the Earth.
- When fossil fuels are burnt, <u>carbon dioxide</u> is produced.
- Although some of this carbon dioxide is removed from the atmosphere when the gas dissolves in the oceans, the overall amount of carbon dioxide in the atmosphere has <u>gradually increased</u> over the last 200 years.
- This carbon dioxide <u>traps the heat</u> that has reached the Earth from the Sun.
- Global warming may mean that the ice at the North and South Poles will melt and cause massive flooding.

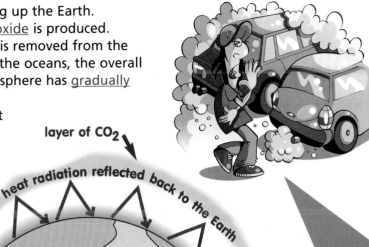

layer of CO_2

heat radiation reflected back to the Earth

light energy from the Sun

Examiner's Top Tip
The greenhouse effect is caused by carbon dioxide.

POLLUTION OF THE ATMOSPHERE

The atmosphere is being polluted in many ways.

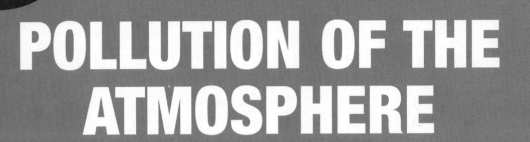

QUICK TEST

1. What is formed when the sulphur in fossil fuels is burnt?

2. What does this form when it dissolves in rain water?

3. What environmental problems can this cause?

4. What gas is formed when fossil fuels are burnt in insufficient oxygen?

5. Why has this gas proved so deadly?

6. What sort of appliances can produce it?

7. What could be the effect of global warming on the environment?

8. Which products contained CFCs?

9. What effect have CFCs had on the ozone layer?

10. Why is the ozone layer important?

Examiner's Top Tip
Remember – acid rain is caused by sulphur dioxide.

10. It filters out harmful UV rays.
9. They break down ozone into oxygen molecules.
8. Aerosols, fridges and solvents
7. Massive flooding
6. Faulty gas appliances
5. It is very poisonous, and difficult to detect because it is colourless and odourless.
4. Carbon monoxide
3. Damage to statues and buildings, trees, animals and plants.
2. Acid rain
1. Sulphur dioxide

PROBLEMS WITH NITRATE FERTILISERS

- Nitrate fertilisers can cause problems if they are washed into streams or lakes.
- Algae (small plants) thrive on the fertilisers and grow extremely well.
- When the algae die, bacteria start breaking down (or decomposing) the algae.
- As the bacteria feed off the algae they use up all the oxygen in the water. Fish and other animals cannot get enough oxygen and die.
- This is called eutrophication.

Nitrate fertilisers can also find their way into our drinking water. There have been concerns over stomach cancer and 'blue baby' disease. Although links have not been proven, it seems wise that their levels should be limited.

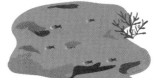

eutrophication

POLLUTION OF THE ENVIRONMENT

OIL SPILLS

Oil is an extremely important material. It is transported around the world in giant oil tankers. However, when accidents occasionally happen crude oil can escape. The oil forms a slick which can devastate animal and plant life.

PROBLEMS WITH LIMESTONE QUARRYING

- Limestone is a very important raw material in industry.
- However, the economic benefits of quarrying for limestone have to be balanced against social and environmental issues.
- Limestone must be blasted from hill-sides in huge quantities which scars the landscape and affects wildlife.
- Transporting the limestone from the quarry can also cause problems.
- However, quarrying also creates new jobs and it brings money into the area.

PROBLEMS DISPOSING OF PLASTIC

- Plastics are <u>unreactive</u>.
- Most do not react with water, oxygen or other chemicals, nor do they get <u>broken down</u> by <u>micro-organisms</u>. This makes them very useful.
- Unfortunately when the plastic is no longer wanted, it does <u>not rot away</u>.
- For this reason plastics are called <u>non-biodegradable</u>.
- They remain in the environment and can cause problems.
- Getting rid of plastics by burning also has problems. Although some plastics burn quite easily, the gases that are given off can be <u>harmful</u>. The plastic PVC releases the gas hydrogen chloride when it is burned.
- In response to these problems some <u>biodegradable plastics</u> have been developed which will <u>rot away</u>.

Examiner's Top Tip
Practise writing these points as 'mini-essay' style answers.

QUICK TEST

1. What are nitrates used for?

2. What will happen to algae if nitrates are present?

3. What happens when the algae dies?

4. What is the name given to this chain of events?

5. Nitrates in drinking water have been linked to which health problems?

6. What are the arguments for and against limestone quarrying?

7. Why are plastics non-biodegradable?

8. Why should plastics not just be burnt?

9. What has been developed in response to these problems?

10. How do oil slicks occur and what effects do they have?

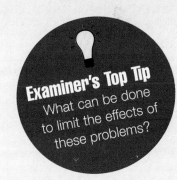

Examiner's Top Tip
What can be done to limit the effects of these problems?

10. Tankers in accidents; oil slicks can devastate animals and plant life.
9. Biodegradable plastics
8. They give off harmful gases.
7. They do not break down in the environment.
6. For: new jobs, money; against: environmental problems.
5. Stomach cancer, 'blue baby' disease
4. Eutrophication
3. The algae is decomposed by bacteria, which uses up oxygen and causes fish, etc to die.
2. It will grow well.
1. Fertilisers

STRUCTURE OF THE EARTH AND PLATE TECTONICS

STRUCTURE OF THE EARTH

- Scientists believe that the Earth has a <u>layered</u> structure.
- The outer layer or <u>crust</u> is very thin. The crust has a low density.
- The next layer down is called the <u>mantle</u>. It extends almost half-way to the centre of the Earth. The rock is mainly solid, but small amounts behave like a viscous fluid and can flow slowly.
- The <u>core</u> has two parts; the outer core is <u>liquid</u> while the inner core, due to the high pressure, is <u>solid</u>.

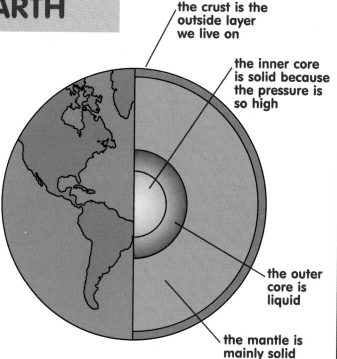

the crust is the outside layer we live on

the inner core is solid because the pressure is so high

the outer core is liquid

the mantle is mainly solid

EVIDENCE OF THE STRUCTURE OF THE EARTH

- Evidence for the layered structure of the Earth comes from examination of the paths of <u>seismic waves</u> (the shock waves sent out by earthquakes). The <u>speed</u> of the waves is affected by the rock they travel through. These studies show that the outer core is <u>liquid</u>.
- The <u>overall</u> density of the Earth is greater than the density of the rocks of the <u>crust</u>.
- This means that the rocks below the crust must be much denser. We believe these rocks are rich in <u>nickel</u> and <u>iron</u>.

Definitions

<u>Core</u> – The core is the central part of the Earth. It is thought to be made of iron and nickel.
<u>Mantle</u> – The mantle is the main bulk of the Earth and is found between the crust and the core.
<u>Crust</u> – The crust is the outermost layer of the lithosphere.
<u>Continental crust</u> – is mainly made of granite. <u>Oceanic crust</u> is mainly basalt.

MOVEMENT OF THE CRUST

HOW THE CONTINENTS ONCE LOOKED

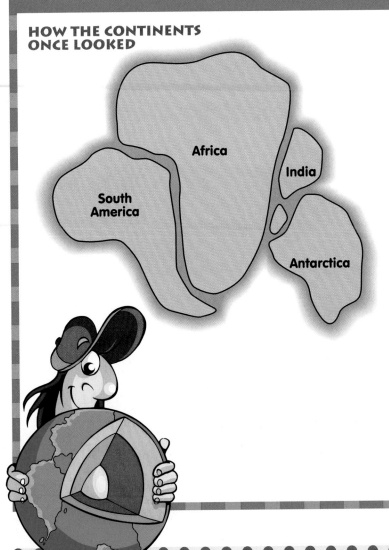

People used to believe that features of the Earth's surface, for example <u>mountain</u> <u>ranges</u>, were formed when the Earth's surface shrank as it cooled.

Scientists now believe that all the geological features of the Earth can be accounted for by a single, unifying theory: plate tectonics.

The main idea in <u>plate</u> <u>tectonics</u> is that the Earth's <u>lithosphere</u>, the <u>crust</u> and the upper part of the <u>mantle</u>, is split into 12 large plates. Each plate moves at a few centimetres per year. The movement is driven by <u>convection</u> <u>currents</u> in the mantle, caused by natural <u>radioactive</u> <u>decay</u> releasing heat.

At one time all the continents were one super continent, called <u>Pangea</u>. Since then the continents have split apart.

QUICK TEST

1. What name is given to the outer layer of the Earth?

2. Which layer is directly below this?

3. What state is this in (solid/liquid/gas)?

4. How do seismic waves give information about the Earth's structure?

5. Why do we believe the core is rich in nickel and iron?

6. How did people used to believe that mountains were formed?

7. What is the name given to the theory which now accounts for geological features?

8. What is the Earth's lithosphere?

9. At what rate do the Earth's plates move?

10. What produces the heat that drives the plates' movements?

10. Radioactive decay
9. A few cm per year
8. The crust and upper part of the mantle.
7. Plate tectonics
6. By the shrinking of the Earth's crust.
5. It is very dense.
4. The speed of the waves is affected as it travels through the rocks.
3. Mainly solid, a little is liquid
2. Mantle
1. Crust

EVIDENCE FOR PLATE TECTONICS

There are many clues that give evidence about plate tectonics:

1. As soon as the American coast was mapped people began to notice how the South American and African coasts <u>fitted</u> <u>together</u> like pieces of a jigsaw.

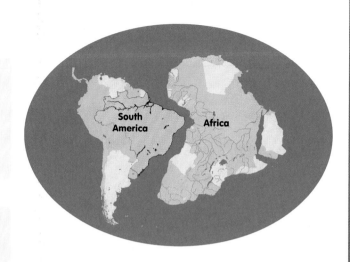

2. Examination of fossil remains in South America and Africa shows that rocks of the <u>same</u> <u>age</u> contained the fossils of a fresh-water crocodile-type creature.

3. More evidence that South America and Africa were once joined was found when <u>rock</u> <u>strata</u> of comparable ages were found to be <u>very</u> <u>similar</u> on both sides of the Atlantic.

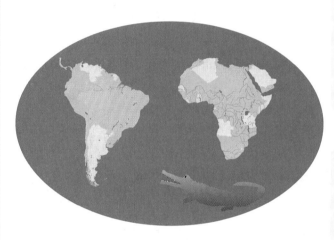

All these pieces of evidence suggest that South America and Africa were <u>once</u> <u>joined</u> <u>together</u> <u>as</u> <u>part</u> <u>of</u> <u>a</u> <u>great</u> <u>land</u> <u>mass</u>. Since that time the continents have split apart.

4. Rocks in Britain from the Carboniferous period (300 million years ago) formed in <u>tropical</u> <u>swamps</u>. 200 million years later Britain was covered by <u>deserts</u>. This shows Britain has moved through <u>different</u> <u>climatic</u> <u>zones</u> as the tectonic plate Britain is on has moved.

This provides further evidence that, over time the <u>Earth's</u> <u>continents</u> <u>have</u> <u>moved</u> <u>around</u> <u>on</u> <u>the</u> <u>surface</u> <u>of</u> <u>the</u> <u>planet</u>.

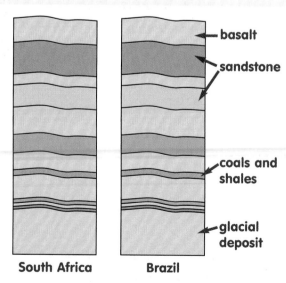

South Africa   Brazil

rock strata from both sides of the Atlantic

- basalt
- sandstone
- coals and shales
- glacial deposit

PLATE BOUNDARIES

This shows how the Earth's <u>lithosphere</u> (crust and upper part of mantle) is <u>cracked</u> <u>into</u> <u>plates</u>.

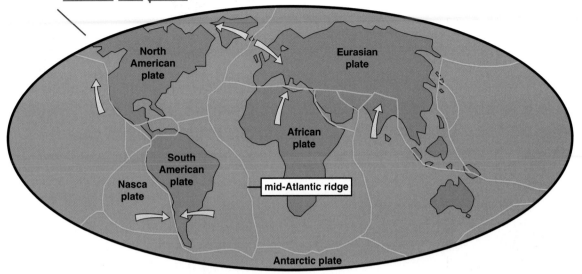

Most problems caused by plate movements <u>occur</u> <u>along</u> <u>the</u> <u>boundaries</u> between plates.

MOVING PLATES

The Earth's lithosphere of the Earth is cracked into a number of plates.

QUICK TEST

1. What is the name of the theory used to explain mountain building?
2. What is strange about the South American and African coast lines?
3. How did the fossil record provide evidence for plate tectonics?
4. How did examination of rock strata provide further evidence?
5. What does the Earth's lithosphere comprise?
6. Where are most of the problems caused by plate movements located?

1. Plate tectonics
2. They fit together.
3. Similar freshwater fossils of a crocodile-type creature were found in both S. America and Africa.
4. Rock strata of comparable ages are similar.
5. Crust and upper mantle
6. Along plate boundaries

PLATE BOUNDARIES

The <u>movement</u> of <u>tectonic plates</u> can cause many problems. These problems are worst near the edges of plates, the plate boundaries, and include <u>earthquakes</u> and <u>volcanoes</u>.

TRANSVERSE/CONSERVATIVE PLATE BOUNDARIES

- Earthquakes are caused by the tectonic plates <u>sliding past</u> each other: The <u>San Andreas fault</u> in California is a famous example.
- The plates in this area are <u>fractured</u> into a complicated pattern.
- As the plates try to move they tend to <u>stick</u> rather than slide smoothly.
- Forces build up on the plates until eventually the strain that has built up is released as an <u>earthquake</u>.
- However with so many factors involved it is <u>not possible to predict</u> exactly when earthquakes will occur.
- When they do occur, they can cause <u>massive destruction and loss of life</u>.

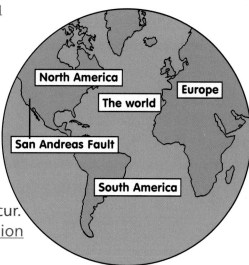

North America

Europe

The world

San Andreas Fault

South America

Damage to gas and water mains caused by an earthquake in Los Angeles in 1994.

CONVERGENT/DESTRUCTIVE PLATE BOUNDARIES

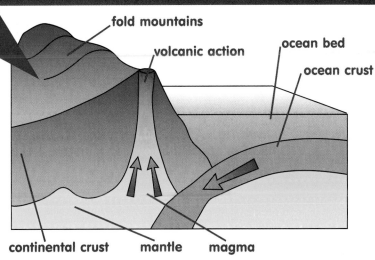

fold mountains
volcanic action
ocean bed
ocean crust
continental crust   mantle   magma

- Convergent plate boundaries often involve the **collision** of an oceanic and a continental plate.

- An **oceanic plate** is **denser** than a continental plate. At a convergent plate boundary between them, the oceanic plate is forced beneath the continental plate.

- The continental crust is **stressed** and folding, and **metamorphism** of existing rock occurs.

- As the oceanic plate is forced down it is **heated** and some of the oceanic plate may melt and form **magma**. This may rise up to create **volcanoes**.

- As the plates are moving past each other **earthquakes** can also occur.

- A convergent plate boundary along the western coast of South America has caused the **Andes mountain range**.

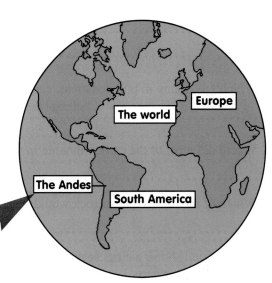

Europe
The world
The Andes
South America

Examiner's Top Tip
Make sure you are able to explain the two types of plate boundary shown here.

QUICK TEST

1. How are earthquakes caused?
2. Where is the San Andreas fault?
3. Why can't the exact time of an earthquake be predicted?
4. Which is denser, an oceanic or a continental plate?
5. At a convergent plate boundary, is the oceanic plate forced above or below the continental plate?
6. How is magma formed at a convergent plate boundary?
7. What would be formed if this magma reachesd the surface?
8. Why are metamorphic rocks found at convergent plate boundaries?
9. Why could earthquakes occur in these areas?
10. Where is the Andes mountain range?

10. West coast of South America
9. Plates move past each other.
8. Heat and pressure changes the existing rocks.
7. A volcano
6. Melting oceanic plate
5. Below
4. Oceanic
3. There are too many factors involved
2. California, USA
1. By plates sliding past each other.

METALS

Three-quarters of the elements in the <u>periodic</u> <u>table</u> are <u>metals</u>. Metals are found in <u>Groups</u> <u>I</u> and <u>II</u> (the left-hand columns) and the transition elements (middle section).

Group	I	II											III	IV	V	VI	VII	0
Period																		
1	H 1																	He 2
2	Li 3	Be 4											B 5	C 6	N 7	O 8	F 9	Ne 10
3	Na 11	Mg 12											Al 13	Si 14	P 15	S 16	Cl 17	Ar 18
4	K 19	Ca 20	Sc 21	Ti 22	V 23	Cr 24	Mn 25	Fe 26	Co 27	Ni 28	Cu 29	Zn 30	Ga 31	Ge 32	As 33	Se 34	Br 35	Kr 36
5	Rb 37	Sr 38	Y 39	Zr 40	Nb 41	Mo 42	Tc 43	Ru 44	Rh 45	Pd 46	Ag 47	Cd 48	In 49	Sn 50	Sb 51	Te 52	I 53	Xe 54
6	Cs 55	Ba 56	57 – 71*	Hf 72	Ta 73	W 74	Re 75	Os 76	Ir 77	Pt 78	Au 79	Hg 80	Tl 81	Pb 82	Bi 83	Po 84	At 85	Rn 86
7	Fr 87	Ra 88	89 – 103**	Rf 104	Db 105	Sg 106	Bh 107	Hs 108	Mt 109	Uun 110	Uuu 111	Uub 112	Uut 113	Uuq 114	Uup 115	Uuh 116	Uus 117	Uuo 118

Key:
- Non-metal
- Metal

PROPERTIES OF METALS

Metals are good conductors of <u>heat</u>.

Metals are good conductors of <u>electricity</u>.

Examiner's Top Tip
Learn the characteristics of metals then cover these pages and write them down.

Metals have high <u>melting</u> and <u>boiling</u> <u>points</u>. All the metals are solids except mercury, which is <u>liquid</u> at room temperature.

Metals are <u>strong</u> and <u>dense</u>, but they are also <u>malleable</u> (can be hammered into shape) and <u>ductile</u> (can be drawn into wires).

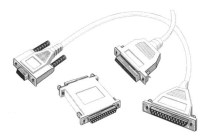

TRANSITION METALS

- **Examples:** <u>iron</u>, <u>nickel</u> and <u>copper</u>
- Transition metals are <u>hard</u> and <u>strong</u>.
- They are much <u>less</u> <u>reactive</u> than the metals in Group I and do not react quickly with oxygen or water.
- Transition metals are widely used:
- Iron is often used as a <u>structural</u> <u>material</u>.
- Copper is a good conductor of both heat and electricity, and it is often used for electrical cables.
- Transition metals form coloured compounds which can be used in pottery glazes.
- Many transition metals and their compounds can act as catalysts. Iron and platinum are widely used in this way.

GROUP I METALS (ALKALI METALS)

Examples: <u>lithium</u>, <u>sodium</u>, <u>potassium</u>

The alkali metals show these properties:
- They have <u>lower</u> <u>densities</u> than typical for metals and <u>float</u> <u>on</u> <u>water</u>;
- They have comparatively <u>low</u> <u>melting</u> <u>points</u> and are quite <u>soft</u> compared with other metals;
- They <u>react</u> <u>vigorously</u> with water, releasing <u>hydrogen</u> <u>gas</u> and forming <u>alkaline</u> <u>solutions</u>.

The Group I metals react with non-metals to form white, ionic compounds which <u>dissolve</u> to form colourless solutions.

Examiner's Top Tip
Remember the uses of transition metals and their compounds.

- -

QUICK TEST

1. Sketch the periodic table and shade Group I and the transition elements.
2. Roughly what fraction of the elements are metals?
3. Name the only metal which is not a solid at room temperature.
4. Name four properties that are common to most metals.
5. Give three examples of alkali metals.
6. Why do alkali metals float on water?
7. Which gas is released when alkali metals react with water?
8. What colour are Group I compounds?
9. Name three transition elements.
10. Give two uses of transition metals.

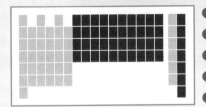

iron/platinum – catalysts
10. Iron – structural; copper – conductors;
9. Iron, nickel, copper etc.
8. White
7. Hydrogen
6. They have low densities.
5. Lithium, sodium, potassium
strong/dense, malleable and ductile
4. Good conductors, high melting points and boiling points,
3. Mercury
2. Three quarters
1. See below

REACTIVITY SERIES

Most reactive	potassium K	
	sodium Na	
	calcium Ca	Extracted from their ores by <u>electrolysis</u>
	magnesium Mg	
	<u>carbon C</u>	
	zinc Zn	
	iron Fe	Extracted from their ores by heating with <u>carbon</u>
	lead Pb	(coke or charcoal)
	<u>hydrogen H</u>	
	copper Cu	Metals less reactive than hydrogen <u>do</u> <u>not</u>
Least reactive	gold Au	react with water or dilute acids

This order has been worked out by observing how vigorous the <u>reaction</u> is between the metal and:

- air
- water
- dilute acid.

REACTIVITY SERIES

Some metals are more reactive than others.
The metals can be placed in order of reactivity.

REACTING METALS WITH AIR

When metals are heated with air they may react with the oxygen present.

metal + *oxygen* $\Rightarrow$ *metal oxide*

magnesium + *oxygen* $\Rightarrow$ *magnesium oxide*

$2Mg(s)$ + $O_2(g)$ $\Rightarrow$ $2MgO(s)$

Most reactive	potassium K	
	sodium Na	
	calcium Ca	These metals react vigorously.
	magnesium Mg	They <u>burn</u> fiercely.
	<u>carbon C</u>	
	zinc Zn	
	iron Fe	These metals react <u>slowly</u> with air.
	lead Pb	
	<u>hydrogen H</u>	
	copper Cu	
Least reactive	gold Au	No reaction

REACTING METALS WITH WATER

Some metals react with water to produce a <u>metal hydroxide</u> and hydrogen.

metal + water ⟹ metal hydroxide + hydrogen

sodium + water ⟹ sodium hydroxide + hydrogen

$2Na(s) + 2H_2O(l)$ ⟹ $2NaOH(aq) + H_2(g)$

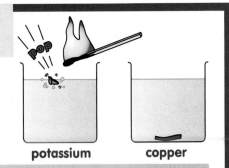

potassium   copper

Most reactive		
	potassium K	
	sodium Na	React <u>vigorously</u> with <u>cold water</u>
	calcium Ca	
	magnesium Mg	
	<u>carbon C</u>	
	zinc Zn	
	iron Fe	React <u>slowly</u> with steam
	lead Pb	
	<u>hydrogen H</u>	
	copper Cu	No reaction
Least reactive	gold Au	

REACTING METALS WITH DILUTE ACIDS

Some metals (those more reactive than hydrogen) react with <u>dilute acids</u> to produce salts and hydrogen.

metal + acid ⟹ salt + hydrogen

calcium + hydrochloric acid ⟹ calcium chloride + hydrogen

$Ca(s) + 2HCl(aq)$ ⟹ $CaCl_2(aq) + H_2(g)$

Most reactive		
	potassium K	
	sodium Na	React <u>violently</u> with dilute acid
	calcium Ca	
	magnesium Mg	
	<u>carbon C</u>	
	zinc Zn	
	iron Fe	<u>Good</u> reaction with dilute acid
	lead Pb	
	<u>hydrogen H</u>	
	copper Cu	No reaction
Least reactive	gold Au	

magnesium   zinc   copper

QUICK TEST

1. How should metals more reactive than carbon be extracted from their ores?
2. How should metals less reactive than carbon be extracted from their ores?
3. How was the reactivity series compiled?
4. When metals burn in oxygen what is formed?
5. What is formed when zinc is burnt in air?
6. Give a balanced equation for this reaction.
7. What is formed when potassium reacts with water?
8. Give a balanced equation for this reaction.
9. What forms when magnesium reacts with hydrochloric acid?
10. Give a balanced equation for this reaction.

10. $Mg(s) + 2HCl(aq) \longrightarrow MgCl_2(aq) + H_2(g)$
9. Magnesium chloride + hydrogen
8. $2K(s) + 2H_2O(l) \longrightarrow 2KOH(aq) + H_2(g)$
7. Potassium hydroxide + hydrogen
6. $2Zn(s) + O_2(g) \longrightarrow 2ZnO(s)$
5. Zinc oxide
4. Metal oxides
air, water, and acid.
3. By observing their reactions with
2. By heating with carbon
1. By electrolysis

METAL DISPLACEMENT REACTIONS

A more reactive metal will displace a less reactive metal from a compound.

REACTIVITY SERIES

Most reactive

potassium K	
sodium Na	
calcium Ca	
magnesium Mg	
	carbon C
zinc Zn	
iron Fe	
lead Pb	
	hydrogen H
copper Cu	
gold Au	

Least reactive

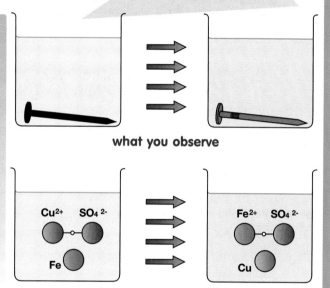

what you observe

what the ions are doing

- Iron is _more reactive_ than copper.

- When an iron nail is placed in a solution of copper sulphate, the nail changes colour from silver to orange-pink.

- The solution changes colour from blue to a very pale green.

- This is an example of a displacement reaction. The more reactive metal, iron, _displaces the less reactive_ metal, copper, from its compound, copper sulphate.

iron + copper sulphate ⇨ copper + iron sulphate
$Fe(s) + CuSO_4(aq)$ ⇨ $Cu(s) + FeSO_4(aq)$

- If the metal which is added is less reactive than the metal in the compound then _no reaction_ will occur.

copper + magnesium sulphate ⇨ no reaction

CORROSION OF METALS

IRON

Iron <u>corrodes</u>, or <u>rusts</u>, faster than most transition metals.
If either oxygen or water is completely removed then iron will <u>not</u> rust.

PREVENTING RUSTING

- **Coating the iron:** painting or coating iron in plastic or oil can stop oxygen and water from reaching it, but if the coating is damaged the iron will rust.

- **Alloying the metal:** if iron is mixed with other metals, such as chromium, it will form the alloy stainless steel. This does not rust.

- **Sacrificial protection:** if a metal which is more reactive than iron, such as zinc or magnesium, is connected to the iron, corrosion will be prevented. Because the zinc is more reactive, the zinc reacts instead of the iron. The iron is protected at the expense of the more reactive metal. For this reason it is called sacrificial protection. Speed-boat engines are protected by attaching a more reactive metal to them.

ALUMINIUM

Some reactive metals like aluminium do not react as quickly as might be suggested by the reactivity series. This is because aluminium reacts with oxygen to form a thin layer of <u>aluminium oxide</u> on the surface of the metal. This stops any more oxygen or water reaching the aluminium, so prevents any further corrosion occurring.

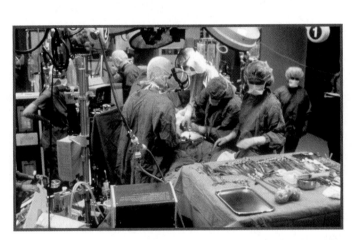

QUICK TEST

1. What is the rule for displacement reactions?
2. What is the word equation for the reaction between magnesium and copper sulphate?
3. Write a balanced symbol equation for the reaction.
4. What is the word equation for the reaction between zinc and iron sulphate?
5. Write a balanced symbol equation for the reaction.
6. What two things are needed for iron to rust?
7. What can be used to coat iron and prevent rusting?
8. How is stainless steel made and what are its advantages?
9. What is 'sacrificial protection'?
10. Why does aluminium not corrode?

1. A more reactive metal will displace a less reactive metal from a compound.
2. Magnesium + copper sulphate —> magnesium sulphate + copper
3. $Mg(s) + CuSO_4(aq) \longrightarrow MgSO_4(aq) + Cu(s)$
4. Zinc + iron sulphate —> zinc sulphate + iron
5. $Zn(s) + FeSO_4(aq) \longrightarrow ZnSO_4(aq) + Fe(s)$
6. Oxygen + water
7. Paint/plastic/oil
8. By alloying iron with another metal such as Chromium, it does not rust.
9. When a more reactive metal e.g. Zinc or Magnesium is put in contact with a less reactive metal e.g. Iron, the less reactive metal is protected as the Zn/Mg reacts first.
10. A thin layer of aluminium oxide prevents any further reaction.

METHODS OF EXTRACTION

The more <u>reactive</u> a metal is, the more difficult it is to remove from its ore. Gold is so unreactive it is found in the Earth's crust on its own (unreacted).

potassium

sodium

calcium   **Metals that are more reactive than carbon are extracted by <u>electrolysis</u>.**

magnesium

aluminium

carbon

zinc

iron   **Metals that are less reactive than carbon are extracted by reducing the metal oxide using <u>carbon</u> (or carbon monoxide).**

tin

lead

gold

Examiner's Top Tip
Learn the equations involved in each of the steps shown.

EXTRACTION OF IRON

Iron is an extremely important metal. It is widely used, particularly once it has been made into steel.

THE BLAST FURNACE

The main ore of iron is iron oxide or haematite Fe_2O_3. Iron is <u>less reactive</u> than carbon so it can be extracted from iron oxide by <u>reduction</u> (removal of oxygen) in the blast furnace.

The solid raw materials added to the blast furnace are:
- **iron ore (haematite)**
- **coke (almost pure carbon)**
- **limestone (reacts with impurities).**

WHAT HAPPENS IN THE BLAST FURNACE?

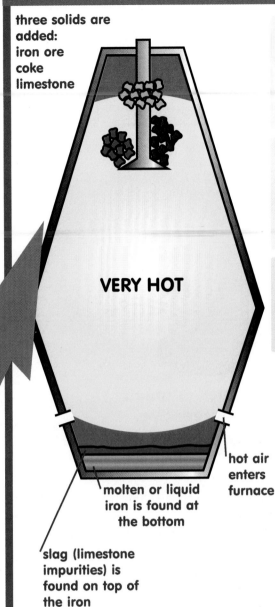

three solids are added:
iron ore
coke
limestone

VERY HOT

molten or liquid iron is found at the bottom

hot air enters furnace

slag (limestone impurities) is found on top of the iron

1. Hot air enters the blast furnace and reacts with carbon, forming carbon dioxide and releasing energy.

carbon + oxygen ⟹ carbon dioxide

$C(s)$ + $O_2(g)$ ⟹ $CO_2(g)$

2. At high temperatures the carbon dioxide reacts with more carbon to form carbon monoxide.

carbon dioxide + carbon ⟹ carbon monoxide

$CO_2(g)$ + $C(s)$ ⟹ $2CO(g)$

3. The carbon monoxide reduces the iron oxide to iron.

carbon monoxide + iron oxide ⟹ iron + carbon dioxide

$3CO(g)$ + $Fe_2O_3(s)$ ⟹ $2Fe(l)$ + $3CO_2(g)$

The iron is dense (heavy for its size) and sinks to the bottom, where it is tapped off.
The carbon monoxide combines with oxygen to form carbon dioxide, so it is oxidised.

REMOVAL OF IMPURITIES

Haematite may contain impurities, most commonly silicon dioxide (silica). When limestone is added it removes the silica by forming molten slag. The slag has a low density so floats on the top of the molten iron ore. The slag can be used in road building or for making fertilisers.

QUICK TEST

1. Which element is so unreactive it can be found uncombined?

2. What method of extraction should be used for metals less reactive than carbon?

3. What method of extraction should be used for metals more reactive than carbon?

4. What is the name and formula of the main iron ore?

5. What three solid raw materials are added to the blast furnace?

6. What other reactant must be added?

7. Which gas actually reduces the iron oxide?

8. Why does the iron sink to the bottom?

9. What substance is formed when limestone reacts with silica?

10. What can this substance be used for?

10. Road building/fertilisers
9. Slag
8. It is more dense.
7. Carbon monoxide
6. Hot air
5. Iron ore, coke, limestone
4. Haematite, Fe_2O_3
3. Electrolysis
2. Heating with carbon
1. Gold

PURIFICATION OF COPPER AND EXTRACTION OF ALUMINIUM

Aluminium is a reactive metal.

Copper is much less reactive.

PURIFICATION OF COPPER

Copper is an <u>unreactive</u> metal, so it can be extracted from its ore by <u>heating</u> with <u>carbon</u>. However, the copper that is produced in this way is not <u>pure</u> enough for use in high-specification electrical wiring. To produce very pure copper, <u>electrolysis</u> is used.

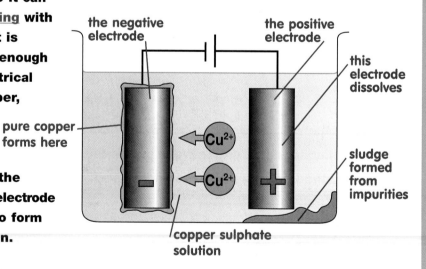

the negative electrode

the positive electrode

this electrode dissolves

pure copper forms here

Cu^{2+}

Cu^{2+}

sludge formed from impurities

copper sulphate solution

WHAT HAPPENS?

1. The impure copper is used as the positive electrode (anode). At this electrode copper atoms give up electrons to form ions which dissolve in the solution.

2. The Cu^{2+} ions are attracted to the negative electrode (cathode).

3. At the negative electrode (cathode) the copper ions gain electrons to form copper atoms. The copper atoms form on the negative electrode, which increases in size.

4. The impurities in the positive electrode fall to the bottom as the positive electrode dissolves away.

EXTRACTION OF ALUMINIUM

Aluminium is <u>more</u> <u>reactive</u> than carbon so it must be extracted by <u>electrolysis</u>, even though this is a very expensive process. The main ore of aluminium is <u>bauxite</u> (aluminium oxide, Al_2O_3). In electrolysis the ions have to be able to move, so the ore has to be <u>dissolved</u> or <u>molten</u>. Bauxite has a very high melting point and heating the ore to this temperature would be very <u>expensive</u>. Fortunately <u>cryolite</u>, another ore of aluminium, has a much lower melting point. Bauxite is <u>dissolved</u> in molten cryolite.

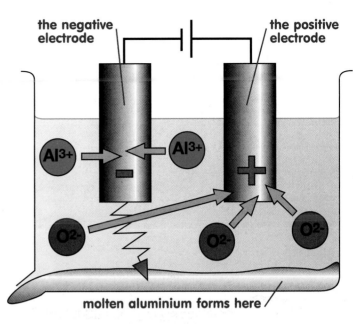

the negative electrode

the positive electrode

Al^{3+} Al^{3+}

O^{2-} O^{2-} O^{2-}

molten aluminium forms here

WHAT HAPPENS?

1. By dissolving the Al_2O_3 both the Al^{3+} and the O^{2-} ions can move.

2. The Al^{3+} ions are attracted to the negative electrode (cathode), where they pick up electrons to form Al atoms. They fall to the bottom of the cell.

3. The O^{2-} ions are attracted to the positive electrode (anode) where they deposit electrons. The oxygen that forms reacts with the graphite electrode, forming carbon dioxide. Periodically the electrodes have to be replaced.

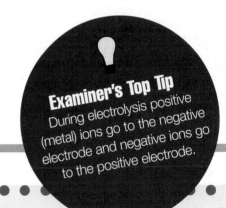

Examiner's Top Tip
During electrolysis positive (metal) ions go to the negative electrode and negative ions go to the positive electrode.

QUICK TEST

1. What is the main ore of aluminium?

2. Why do the ions have to be dissolved or molten?

3. Which other ore of aluminium is used in electrolysis?

4. Why is it used?

5. During electrolysis what forms at the positive electrode?

6. What is formed at the negative electrode?

7. How is copper normally extracted from its ore?

8. Why is very pure copper sometimes needed?

9. At which electrode is the impure copper placed?

10. What happens to any impurities?

10. They form a sludge at the bottom of the cell.
9. Positive electrode
8. High-specification wiring
7. Normally heated with carbon
6. Aluminium
5. Oxygen, which reacts to make carbon dioxide
4. It has a lower melting point than bauxite, so the bauxite dissolves in it.
3. Cryolite
2. So they can move
1. Bauxite

INDICATORS

- The <u>pH</u> shows the concentration of hydrogen ions [H$^+$] in a solution.
- Indicators show whether a solution is acidic, alkaline or neutral by changing colour.
- There are many different indicators.

Indicator	Acid	Neutral	Alkali
Universal Indicator	red	green	purple
Blue litmus	red	blue	blue
Red litmus	red	red	blue
Phenolphthalein	colourless	colourless	pink

ACIDS AND ALKALIS

ACIDS

<u>Acidic</u> solutions have a pH less than 7.
The <u>strongest</u> acids have a pH of 1.
The <u>weakest</u> acids have a pH of 6.

COMMON ACIDS ARE:

- hydrochloric acid
- sulphuric acid
- nitric acid

The soluble oxides of non-metals form acidic solutions.

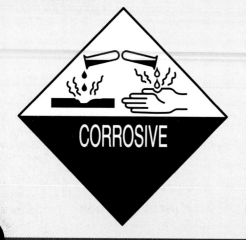

CORROSIVE

ALKALIS

Alkalis are also corrosive.
Alkalis are also called bases.
Alkalis are soluble bases.
Alkalis have a pH of more than 7.
The strongest alkalis have a pH of 14.
The weakest alkalis have a pH of 8.

COMMON ALKALIS ARE:

- sodium hydroxide
- potassium hydroxide
- calcium hydroxide

Ammonia dissolves in water to form an alkaline solution. This can be neutralised with acids to produce ammonium salts.

The soluble oxides and hydroxides of metals form alkaline solutions.

Examiner's Top Tip
Hydrogen, H$^+$ ions make solutions acidic.

NEUTRALISATION

The reaction between an acid and an alkali is called <u>neutralisation</u>.
acid + alkali ⇨ a neutral salt + water

The type of salt produced depends on the metal in the alkali used and on the acid used.

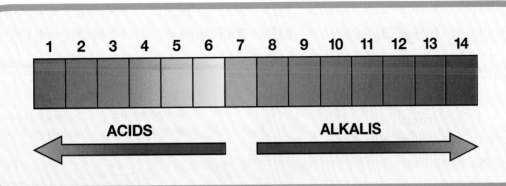

ACIDS   ALKALIS

NAMING SALTS

Neutralising <u>hydrochloric</u> acid will produce <u>chloride</u> salts:
- hydro<u>chloric</u> acid + <u>sodium</u> hydroxide ⇨ <u>sodium</u> <u>chloride</u> + water

Neutralising <u>nitric</u> acid will produce <u>nitrate</u> salts:
- <u>nitric</u> acid + <u>calcium</u> hydroxide ⇨ <u>calcium</u> <u>nitrate</u> + water

Neutralising <u>sulphuric</u> acid will produce <u>sulphate</u> salts:
- <u>sulphuric</u> acid + <u>potassium</u> hydroxide ⇨ <u>potassium</u> <u>sulphate</u> + water

Examiner's Top Tip
Hydroxide, [OH⁻] ions make solutions alkaline.

QUICK TEST

1. What is the pH of a neutral solution?
2. What is the pH of the strongest alkali?
3. What is the pH of a weak acid?
4. Which ions make solutions acidic?
5. Which ions make solutions alkaline?
6. How are alkalis and bases related?
7. Name three common acids.
8. Name the salt produced when sulphuric acid neutralises sodium hydroxide.
9. Name the salt produced when nitric acid neutralises potassium hydroxide.
10. Name the salt produced when hydrochloric acid neutralises ammonia solution.

10. Ammonium chloride
9. Potassium nitrate
8. Sodium sulphate
7. Hydrochloric acid, sulphuric acid, nitric acid
6. Alkalis are soluble bases.
5. OH⁻
4. H⁺
3. 6
2. 14
1. 7

METAL CARBONATES

Metal carbonates can be <u>neutralised</u> by acids.

Most carbonates are <u>insoluble</u>, so they are bases, but they are not alkalis.

When carbonates are neutralised carbon dioxide is given off:

<u>metal</u> <u>carbonate</u> + <u>acid</u> ⟹ <u>salt</u> + <u>water</u> + <u>carbon dioxide</u>

copper carbonate + hydrochloric acid ⟹ copper chloride + water + carbon dioxide

$CuCO_3(s)$ + $2HCl(aq)$ ⟹ $CuCl_2(aq)$ + $H_2O(l)$ + $CO_2(g)$

zinc carbonate + sulphuric acid ⟹ zinc sulphate + water + carbon dioxide

$ZnCO_3(s)$ + $H_2SO_4(aq)$ ⟹ $ZnSO_4(aq)$ + $H_2O(l)$ + $CO_2(g)$

MAKING COPPER CHLORIDE

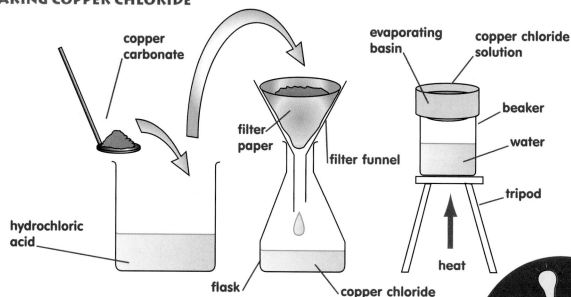

copper carbonate

filter paper

filter funnel

evaporating basin

copper chloride solution

beaker

water

tripod

heat

hydrochloric acid

flask

copper chloride solution

- *Copper carbonate is added to the acid until it stops fizzing.*
- *The unreacted copper carbonate is then removed by <u>filtering</u>.*
- *The solution is poured into an evaporating dish.*
- *It is heated until the first crystals appear.*
- *The solution is then left for a few days for the copper chloride to <u>crystallize</u>.*

Examiner's Top Tip
Sulphuric acid makes sulphate salts.
Hydrochloric acid makes chloride salts.
Nitric acid makes nitrate salts.

METALS

Metals can be reacted with acids to form a salt and hydrogen:

• <u>metal</u> + <u>acid</u> ⟹ <u>salt</u> + <u>hydrogen</u>

zinc + hydrochloric acid ⟹ zinc chloride + hydrogen

$Zn(s) + 2HCl(aq)$ ⟹ $ZnCl_2(aq)$ + $H_2(g)$

magnesium + sulphuric acid ⟹ magnesium sulphate + hydrogen

$Mg(s)$ + $H_2SO_4(aq)$ ⟹ $MgSO_4(aq)$ + $H_2(g)$

METAL OXIDES

Metal oxides are also bases; they can be reacted with acids to make salts and water:

- **metal oxide** + **acid** ⟹ **salt** + **water**

copper oxide + hydrochloric acid ⟹ copper chloride + water

$CuO(s)$ + $2HCl(aq)$ ⟹ $CuCl_2(aq)$ + $H_2O(l)$

zinc oxide + sulphuric acid ⟹ zinc sulphate + water

$ZnO(s)$ + $H_2SO_4(aq)$ ⟹ $ZnSO_4(aq)$ + $H_2O(l)$

MAKING SALTS

METAL HYDROXIDES

We have seen that metal hydroxides can be neutralised with acids to make salt and water:

metal hydroxide + acid ⟹ salt + water

QUICK TEST

1. What is formed when hydrochloric acid reacts with potassium hydroxide?
2. What is formed when sulphuric acid reacts with sodium hydroxide?
3. Which gas is given off when carbonates react with acid?
4. What is formed when hydrochloric acid reacts with zinc carbonate?
5. What is formed when sulphuric acid reacts with magnesium carbonate?
6. How could you get a sample of a soluble salt?
7. What is formed when hydrochloric acid reacts with magnesium?
8. What is formed when sulphuric acid reacts with zinc?
9. What is formed when hydrochloric acid reacts with zinc oxide?
10. What is formed when sulphuric acid reacts with copper oxide?

1. Potassium chloride + water
2. Sodium sulphate + water
3. Carbon dioxide
4. Zinc chloride + water + carbon dioxide
5. Magnesium sulphate + water + carbon dioxide
6. Remove unreacted solid by filtering, then evaporate off the water.
7. Magnesium chloride + hydrogen
8. Zinc sulphate + hydrogen
9. Zinc chloride + water
10. Copper sulphate + water

1. Four metals were placed in hydrochloric acid; the diagram shows what happened.
   Place the metals in order of reactivity (most reactive first).

....

Iron   Zinc   Copper   Magnesium

2. Give an example of a strong acid.

....

3. Give an example of a strong alkali.

....

4. What substance could you use to find the pH of a colourless solution?

....

5. A solution has a pH between 5 and 6. Describe what you would see if this solution was tested with blue litmus paper.

....

6. Why is copper less likely to corrode than iron?

....

7. What colour would universal indicator be in a solution of pH 1?

....

8. Put these metals in order of reactivity, most reactive first:
   gold   sodium   iron   zinc

....

9. Copper can be extracted from copper sulphate solution by adding an iron nail. Why does the copper sulphate react with the iron?

....

10. Iron is extracted in a blast furnace. Which solid substance is added to the blast furnace along with iron ore and limestone?

....

11. Name the main ore of aluminium.

....

12. Name the type of reaction when an alkali is added to an acid.

....

13. Name the process by which aluminium is extracted from its ore.

....

14. During electrolysis, why do the ions have to be molten or dissolved in water?

....

15. Which gas actually reduces the iron oxide to iron in the blast furnace?

....

16. Name the salt produced when:
    a) hydrochloric acid reacts with calcium hydroxide

....

    b) sulphuric acid reacts with potassium hydroxide

....

    c) nitric acid reacts with sodium hydroxide

....

17. Name the two products formed when sodium reacts with water.

....

18. In a displacement reaction a more reactive metal takes the place of a less reactive metal in its compound. The metals are listed below in order of reactivity:

most reactive: magnesium
                     iron
                     copper

The three metals are added to solutions of metal sulphates as shown below. A tick shows a reaction; a cross shows no reaction. There is a reaction between magnesium and copper sulphate which is shown by a tick. Complete the table by adding a tick to show a reaction or a cross to show no reaction:

metal	solution		
	magnesium sulphate	copper sulphate	iron sulphate
magnesium	✕	✔	
copper		✕	
iron			✕

19. A newly discovered metal is more reactive than carbon. How should the metal be extracted from its ore?

....

....

20. Wasp stings can be treated with vinegar; bee stings can be treated with dilute ammonia.
    a) Are wasp stings acidic or alkaline?

....

....

    b) Are bee stings acidic or alkaline?

....

....

How did you do?

1–5	correct ....start again
6–10	correct ....getting there
11–15	correct ....good work
16–20	correct ....excellent

SOLIDS

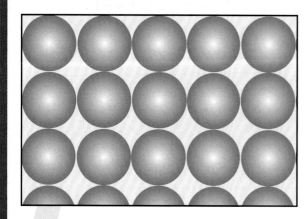

- Particles <u>are</u> <u>very</u> <u>close</u> <u>together</u>
- Particles <u>are</u> <u>held</u> <u>together</u> by <u>strong</u> <u>forces</u> of <u>attraction</u>
- Particles <u>vibrate</u> but have <u>fixed</u> <u>positions.</u>

KEY POINT
➡ Solids have a <u>definite</u> <u>shape</u> and <u>volume</u> and are <u>hard</u> <u>to</u> <u>compress.</u>

STATES OF MATTER

THERE ARE ❸ <u>STATES</u> <u>OF</u> <u>MATTER</u>: <u>SOLID</u>, <u>LIQUID</u> AND <u>GAS</u>.

LIQUIDS

- <u>Particles</u> <u>are</u> <u>close</u> <u>together</u>
- <u>Particles</u> <u>are</u> <u>held</u> <u>together</u> by <u>forces</u> <u>of</u> <u>attraction</u>
- <u>Particles</u> <u>move</u> <u>relative</u> <u>to</u> each other.

KEY POINT
➡ Liquids have a <u>definite</u> <u>volume</u>, but <u>not</u> a definite shape and are <u>hard</u> <u>to</u> <u>compress.</u>

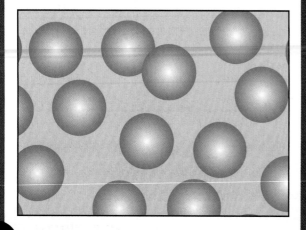

GASES

- <u>Particles</u> <u>are</u> <u>far</u> <u>apart</u> <u>from</u> each other
- There are <u>no</u> <u>forces</u> <u>of</u> <u>attraction</u> between particles
- <u>Particles</u> <u>move</u> relative to each other.

KEY POINT
➡ <u>Gases</u> do <u>not</u> have a <u>definite</u> <u>shape</u> or <u>volume</u> and are <u>easy</u> <u>to</u> <u>compress.</u>

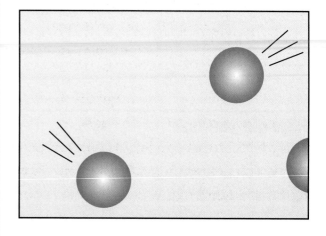

CHANGES OF STATE

When changing from one state to another there is no change in mass.

solid (ice) → liquid (water)

MELTING AND BOILING POINTS

- Above its boiling point a substance is a <u>gas</u>.
- Between its melting point and its boiling point a substance is a <u>liquid</u>.
- Below its melting point a substance is a <u>solid</u>.

- At 25°C (room temperature) <u>oxygen</u> is a <u>gas</u>. 25°C is above the boiling point of oxygen.
- At 25°C <u>mercury</u> is a <u>liquid</u>. 25°C is above the melting point but below the boiling point of mercury.
- At 25°C <u>iron</u> is a <u>solid</u>. 25°C is below the melting point of iron.

substance	melting point (°C)	boiling point (°C)
iron	1535	2750
mercury	–39	357
oxygen	–218	–183

CHANGING STATES

temperature increases

a) The particles of the solid are heated and vibrate more.

b) The vibration of the particles overcomes the forces of attraction between the particles.

c) The particles of the liquid are heated and move more quickly.

d) The movement of the liquid particles overcomes the forces of attraction between the particles.

e) The particles in the gas move faster.

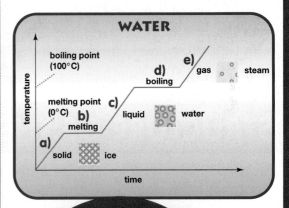

Examiner's Top Tip
This is quite basic stuff so make sure you are really familiar with it. Draw a temperature-time graph to show steam condensing to form liquid water.

QUICK TEST

1. Name the three states of matter.
2. In which of the states are the particles closest together?
3. Do solids have a definite volume?
4. In which state are particles held together by forces of attraction, but the particles may move relative to each other?
5. Can liquids be compressed?
6. Are there any forces of attraction between gas particles?
7. Can gases be compressed easily?
8. In which process do liquids turn into gases?
9. In which process do solids turn into liquids?
10. Draw a temperature-time graph to show solid ice melting to form liquid water.

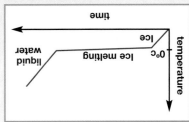

1. Solid, liquid and gas 2. Solid 3. Yes 4. Liquid 5. No 6. No 7. Yes 8. Boiling 9. Melting 10. See below

ATOMIC STRUCTURE

An atom has a nucleus surrounded by shells of electrons.

The <u>electrons</u> are found in shells around the <u>nucleus</u>.

The <u>nucleus</u> is found at the centre of the <u>atom</u> and contains <u>neutrons</u> and <u>protons</u>.

STRUCTURE OF THE ATOM

Protons have a <u>positive</u> <u>charge</u> and a mass of 1.
Neutrons have <u>no</u> <u>charge</u> and have a mass of 1.
Electrons have a <u>negative</u> <u>charge</u> and a negligible mass.

Particle	Mass	Charge
Protons	1	+1
Neutrons	1	0
Electrons	negligible	-1

Examiner's Top Tip
Ensure you are familiar with the charge and mass of the three types of particles.

Examiner's Top Tip
In chemical reactions only the electrons are involved.

In all neutral atoms there is <u>no overall charge</u>, so the number of protons is equal to the number of electrons.

$$^{23}_{11}\text{Na}$$

The <u>mass number</u> is the number of protons added to the number of neutrons.
The <u>atomic number</u> is the number of protons.

- Sodium has an atomic number of 11, so it has 11 protons.
- The sodium atom has <u>no</u> overall charge so the number of electrons must be the same as the number of protons. Sodium therefore has 11 electrons.
- The number of neutrons is given by the mass number minus the atomic number.
- Sodium has 23 – 11 = 12 neutrons.

ISOTOPES

Isotopes of an element have the <u>same</u> number of protons but a <u>different</u> number of neutrons. So isotopes have the same <u>atomic</u> <u>number</u> but a different <u>mass</u> <u>number</u>.

• Chlorine has 2 common isotopes:

$^{35}_{17}Cl$
- 17 protons
- 17 electrons
- <u>18</u> <u>neutrons</u>

$^{37}_{17}Cl$
- 17 protons
- 17 electrons
- <u>20</u> <u>neutrons</u>

• The isotopes will react chemically in the same way because they have identical numbers of electrons.

THE ATOMIC STRUCTURE OF THE FIRST 20 ELEMENTS

SYMBOL	ELEMENTS	PROTONS	ELECTRONS	NEUTRONS	SYMBOL	ELEMENTS	PROTONS	ELECTRONS	NEUTRONS
$^{1}_{1}H$	Hydrogen	1	1	0	$^{23}_{11}Na$	Sodium	11	11	12
$^{4}_{2}He$	Helium	2	2	2	$^{24}_{12}Mg$	Magnesium	12	12	12
$^{7}_{3}Li$	Lithium	3	3	4	$^{27}_{13}Al$	Aluminium	13	13	14
$^{9}_{4}Be$	Beryllium	4	4	5	$^{28}_{14}Si$	Silicon	14	14	14
$^{11}_{5}B$	Boron	5	5	6	$^{31}_{15}P$	Phosphorus	15	15	16
$^{12}_{6}C$	Carbon	6	6	6	$^{32}_{16}S$	Sulphur	16	16	16
$^{14}_{7}N$	Nitrogen	7	7	7	$^{35}_{17}Cl$	Chlorine	17	17	18
$^{16}_{8}O$	Oxygen	8	8	8	$^{40}_{18}Ar$	Argon	18	18	22
$^{19}_{9}F$	Fluorine	9	9	10	$^{39}_{19}K$	Potassium	19	19	20
$^{20}_{10}Ne$	Neon	10	10	10	$^{40}_{20}Ca$	Calcium	20	20	20

QUICK TEST

1. What does the nucleus contain?

2. What are found in shells around the nucleus?

3. What is the charge and mass of a proton?

4. What is the charge and mass of an electron?

5. What is the charge and mass of a neutron?

6. What is the mass number of an atom?

7. What is the atomic number of an atom?

8. What is the same about the atoms of two isotopes of an element?

9. What is different about the atoms of two isotopes of an element?

10. Why do isotopes of an element react in the same way?

Examiner's Top Tip
Be able to explain what isotopes are.

10. They have the same number of electrons.
9. Mass number/number of neutrons
8. Atomic number/number of protons or electrons
7. Number of protons
6. Number of protons + number of neutrons
5. No charge, mass 1
4. Charge −1, mass negligible
3. Charge +1, mass 1
2. Electrons
1. Protons and neutrons

ELECTRON STRUCTURE

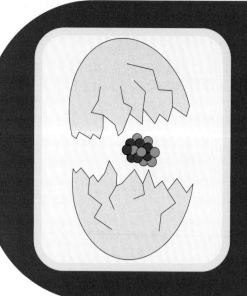

The electrons are found in shells around the nucleus.

ELECTRON STRUCTURE

- Electrons occupy the lowest available shell, this is the one <u>closest</u> to the nucleus.
- The first shell may contain up to <u>two</u> electrons.
- The second and third shells may contain up to <u>eight</u> electrons.
- The number of electrons in the outer shell indicates the group that the element belongs to. So, the electron structure shows how the atom will react chemically.

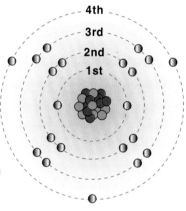

a model of electron shells

LITHIUM
- Number of protons = 3
- Number of electrons = 3
- The electron structure is 2, 1.

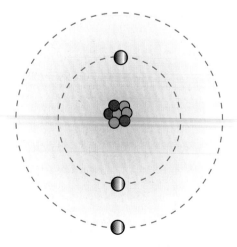

lithium is in <u>Group I</u>

MAGNESIUM
- Number of protons = 12
- Number of electrons = 12
- The electron structure is 2, 8, 2.

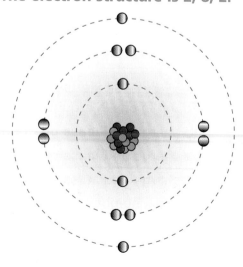

magnesium is in <u>Group II</u>

THE ELECTRON STRUCTURE OF THE FIRST 20 ELEMENTS

Symbol	Element	Electrons	Electron Structure	Group
H	Hydrogen	1	1	I
He	Helium	2	2	II
Li	Lithium	3	2, 1	I
Be	Beryllium	4	2, 2	II
B	Boron	5	2, 3	III
C	Carbon	6	2, 4	IV
N	Nitrogen	7	2, 5	V
O	Oxygen	8	2, 6	VI
F	Fluorine	9	2, 7	VII
Ne	Neon	10	2, 8	O
Na	Sodium	11	2, 8, 1	I
Mg	Magnesium	12	2, 8, 2	II
Al	Aluminium	13	2, 8, 3	III
Si	Silicon	14	2, 8, 4	IV
P	Phosphorus	15	2, 8, 5	V
S	Sulphur	16	2, 8, 6	VI
Cl	Chlorine	17	2, 8, 7	VII
Ar	Argon	18	2, 8, 8	O
K	Potassium	19	2, 8, 8, 1	I
Ca	Calcium	20	2, 8, 8, 2	II

QUICK TEST

1. Where are electrons found?
2. How many electrons may go into the first shell?
3. In this model, how many electrons may go into the second and third shells?
4. If an atom has 7 protons, how many electrons will it have?
5. If an atom has 11 electrons, what is its electron structure?
6. Another atom has 17 electrons. What is its electron structure and to which group does it belong?

6. 2, 8, 7; Group VII
5. 2, 8, 1
4. 7
3. Up to 8
2. Up to 2
1. Around the nucleus in the lowest available shell.

IONIC BONDING

Ionic bonding involves the transfer of electrons.

IONIC BONDING

All atoms wish to have a <u>full outer shell</u> of electrons (like the noble gases).
Ionic bonding involves the <u>transfer</u> of electrons from one atom to another.
Metals in Groups I and II – such as sodium and calcium – lose negative electrons to gain
a full outer shell. Overall they become positively charged (electrons are negative).
Non-metals in Groups VI and VII – such as oxygen and chlorine – gain negative electrons
to attain a full outer shell. So overall they become negatively charged.
The table shows the ions formed by the first 20 elements.

H+ hydrogen								Group 0 None helium
Group I	**Group II**		**Group III**	**Group IV**	**Group V**	**Group VI**	**Group VII**	
Li+ lithium	Be²⁺ beryllium		None boron	None carbon	None nitrogen	O²⁻ oxide	F⁻ fluoride	None neon
Na+ sodium	Mg²⁺ magnesium		Al³⁺ aluminium	None silicon	None phosphorus	S²⁻ sulphide	Cl⁻ chloride	None argon
K+ potassium	Ca²⁺ calcium	transition metals						

EXAMPLE

In neutral atoms there are the <u>same number</u> of <u>positive protons</u> as
there are <u>negative electrons</u>. This means that if electrons are lost
or gained the number of <u>protons</u> and the number of <u>electrons</u>
is no longer <u>balanced</u>. So an ion is an <u>atom</u>, or a <u>small group</u> of
atoms, with a <u>charge</u>.

sodium + chlorine ⟹ sodium chloride

The sodium atom <u>transfers</u> an electron to the chlorine atom. Both the
sodium and the chlorine atoms now have full outer shells. Sodium has lost
a negative electron so becomes <u>positively</u> charged. Chlorine has <u>gained</u> an
electron so becomes <u>negatively</u> charged.

sodium + fluorine ⟹ sodium fluoride

Fluorine and chlorine are in the same group of the periodic table,
Group VII. Fluorine and sodium will therefore react in a similar
way to chlorine and sodium. Sodium transfers
one electron to fluorine.

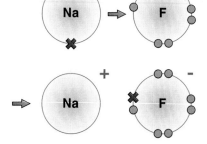

IONIC BONDING II

The dots and crosses represent electrons.

EXAMPLES

magnesium + oxygen $\Rightarrow$ magnesium oxide

Mg + $\frac{1}{2}O_2$ $\Rightarrow$ Mg^{2+} + O^{2-}

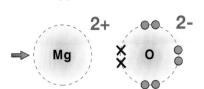
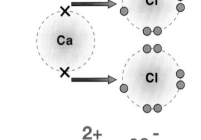

- Magnesium transfers two electrons to oxygen.

calcium + chlorine $\Rightarrow$ calcium chloride

Ca + Cl_2 $\Rightarrow$ Ca^{2+} + Cl^- Cl^-

- Calcium transfers two electrons in total, one to each chlorine atom.

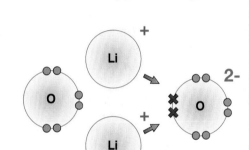

lithium + oxygen $\Rightarrow$ lithium oxide

Li Li + O_2 $\Rightarrow$ Li^+ Li^+ + O^{2-}

- Each lithium atom transfers one electron to the oxygen atom.

All these compounds are held together by strong forces of attraction between the oppositely charged ions.

***

QUICK TEST

1. If an element in Group I loses an electron, what charge does it have?
2. If an element in Group VI gains two electrons, what charge does it have?
3. Draw a dot and cross diagram to show sodium reacting with chlorine.
4. Draw a dot and cross diagram to show calcium reacting with chlorine.
5. What holds together ionic compounds?

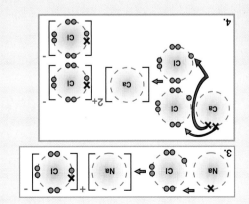

5. Strong attraction between oppositely charged ions
3. & 4. See below
2. 2−
1. 1+

COVALENT BONDING

This involves the <u>sharing of electrons</u>.

HOW IT WORKS

Covalent bonding occurs between <u>non-metals</u>. The atoms share electrons in the bond. Covalent bonding allows both atoms to feel that they have a <u>stable</u>, full outer <u>shell</u>.

HYDROGEN, H_2

Both hydrogen atoms have only one electron, but by forming a single covalent bond, both can have a full outer shell.

This can also be shown as   H – H

HYDROGEN CHLORIDE, HCl

The hydrogen and the chlorine atoms both need <u>one</u> more electron. They form a single covalent bond so both have a full outer shell.

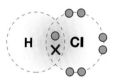

This can also be shown as   H – Cl

METHANE, CH_4

The carbon has four outer electrons so needs <u>four</u> more for a full outer shell. The carbon forms four single covalent bonds to the hydrogen atoms, so all the atoms now have a full outer shell of electrons.

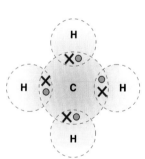

This can also be shown as

$$\begin{array}{c} H \\ | \\ H - C - H \\ | \\ H \end{array}$$

AMMONIA, NH_3

The nitrogen atom has <u>five</u> outer electrons so needs <u>three</u> more. Nitrogen forms three single covalent bonds to hydrogen atoms.

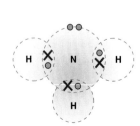

This can also be shown as

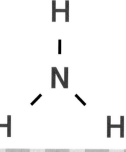

MORE EXAMPLES

Examiner's Top Tip
This is an important section make sure you learn it really well.

WATER, H_2O

The oxygen has <u>six</u> outer electrons so needs <u>two</u> more. The oxygen forms two single covalent bonds with the two hydrogen atoms to give it a full outer shell.

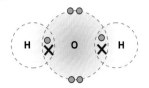

This can also be shown as

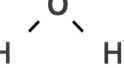

OXYGEN, O_2

Both oxygen atoms have six outer electrons so both need two more. The oxygen atoms form one double covalent bond so that both have a full outer shell.

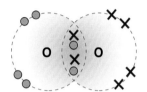

This can also be shown as $O = O$

CARBON DIOXIDE, CO_2

The carbon has four outer electrons, so needs four more. It forms double covalent bonds with two oxygen atoms, so that all the atoms now have a full outer shell of electrons.

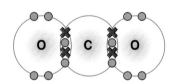

This can also be shown as $O = C = O$

QUICK TEST

1. How many outer electrons do elements in Group I have?
2. How many outer electrons do elements in Group VI have?
3. What sort of bonding occurs between non-metal atoms?
4. Draw a dot and cross diagram to show hydrogen bonding with chlorine.
5. Draw a dot and cross diagram to show the bonding in methane.

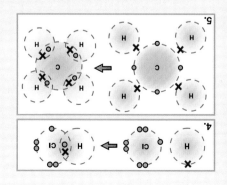

IONIC AND COVALENT COMPOUNDS

Ionic bonding occurs between <u>metals</u> in Groups I and II and <u>non-metals</u> in Groups VI and VII. It involves the transfer of electrons. Covalent bonding occurs between non-metal atoms. It involves the sharing of electrons.

IONIC COMPOUNDS

- *Ionic compounds are held together by the strong forces of attraction between oppositely charged ions.*
- *The compound has a regular structure.*
- *Ions form giant structures.*
- *The strong forces of attraction between oppositely charged ions means they have very high melting and boiling points.*
- *Ionic compounds can often be <u>dissolved</u> in <u>water</u>.*

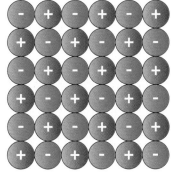

- *When dissolved in water to form solutions the ions can move, so they are able to conduct electricity.*
- *Similarly, if ionic compounds are heated until they melt, the ions can move and they can conduct electricity.*

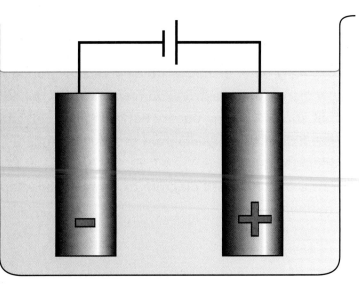

When molten or dissolved in water, ionic compounds can conduct electricity

COVALENT STRUCTURES

Atoms which <u>share</u> <u>electrons</u> can form molecules. The atoms are <u>held</u> <u>together</u> <u>by</u> <u>shared</u> <u>pairs</u> <u>of</u> <u>electrons</u>.

SIMPLE MOLECULAR COVALENT STRUCTURES
Examples
- Oxygen, water, ammonia, methane and carbon dioxide

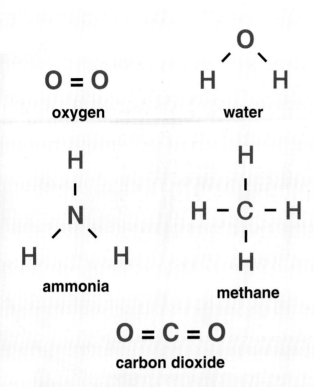

oxygen

water

ammonia

methane

carbon dioxide

one way to represent covalent bonds

These molecules are formed from small numbers of atoms.
There are strong covalent bonds between the atoms in each molecule, but very weak forces of attraction between these molecules.
This means that simple molecular compounds have low melting and boiling points.
Most are gases or liquids at room temperature.
Simple molecular substances <u>do</u> <u>not</u> <u>conduct</u> <u>electricity</u> because they do not contain ions.
They tend to be <u>insoluble</u> in water (although they may dissolve in other solvents).

Examiner's Top Tip
Questions on these areas come up very frequently, make sure you know this section really well.

QUICK TEST

1. What kind of structure do ionic compounds form?

2. Why do they have high melting and boiling points?

3. Why can ionic substances conduct electricity when dissolved but not when they are solid?

4. Why can ionic substances conduct when they are molten?

5. Give an example of a simple molecular substance.

6. In simple molecular compounds describe the bonding between atoms and between molecules.

7. Can simple molecular compounds ever conduct electricity?

7. No
6. Strong attraction between atoms, weak between molecules
5 Oxygen/water/ammonia/methane/carbon dioxide, etc.
4. Ions can move when they are molten.
3. When dissolved the ions can move
2. Because of the strong forces of attraction between oppositely charged ions.
1. Giant and regular structures

THE PERIODIC TABLE

As the elements were discovered, early chemists tried to find <u>patterns</u> amongst them, but they struggled to find <u>links</u>.

Group	I	II											III	IV	V	VI	VII	0
Period																		
1	H 1																	He 2
2	Li 3	Be 4											B 5	C 6	N 7	O 8	F 9	Ne 10
3	Na 11	Mg 12											Al 13	Si 14	P 15	S 16	Cl 17	Ar 18
4	K 19	Ca 20	Sc 21	Ti 22	V 23	Cr 24	Mn 25	Fe 26	Co 27	Ni 28	Cu 29	Zn 30	Ga 31	Ge 32	As 33	Se 34	Br 35	Kr 36
5	Rb 37	Sr 38	Y 39	Zr 40	Nb 41	Mo 42	Tc 43	Ru 44	Rh 45	Pd 46	Ag 47	Cd 48	In 49	Sn 50	Sb 51	Te 52	I 53	Xe 54
6	Cs 55	Ba 56	57 – 71*	Hf 72	Ta 73	W 74	Re 75	Os 76	Ir 77	Pt 78	Au 79	Hg 80	Tl 81	Pb 82	Bi 83	Po 84	At 85	Rn 86
7	Fr 87	Ra 88	89 – 103**	Rf 104	Db 105	Sg 106	Bh 107	Hs 108	Mt 109	Uun 110	Uuu 111	Uub 112	Uut 113	Uuq 114	Uup 115	Uuh 116	Uus 117	Uuo 118

Note that elements 113, 115 and 117 are not yet known, but are included in the table to show their respective positions. Elements 114, 116 and 118 have only been reported recently.

THE PERIODIC TABLE

- In the modern periodic table the elements are arranged in order of <u>increasing</u> atomic number.
- The elements are placed in <u>rows</u> so that elements with <u>similar</u> <u>properties</u> are in the <u>same</u> <u>column</u>.
- These vertical columns are called <u>groups</u>.
- They are often numbered using roman numerals, for example <u>Group I</u>, which consists of Li, Na, K, Rb, Cs, Fr.
- All the members of Group I share similar properties: they are all metals which react to form ions with a <u>1+ charge</u>.
- All the elements in Group I have one <u>electron</u> in their outer shell.
- The horizontal rows in the periodic table are called <u>periods</u>. Across a period an electron shell is gradually filled with electrons. In the next period the next electron shell is gradually filled.

THE HISTORY

1864 JOHN NEWLANDS

<u>Newlands</u> arranged the known elements in rows of seven, according to their <u>atomic mass</u>.

Li	Be	B	C	N	O	F
Na	Mg	Al	Si	P	S	Cl

Newlands noticed <u>similarities</u> between every eighth element (the noble gases were not discovered until later).
Newlands had identified <u>periodicity</u>, but because he left no gaps for the elements yet to be discovered, many problems developed.

1869 DIMITRI MENDELEEV

<u>Mendeleev</u> realised that some elements had not yet been discovered. He ordered the elements by their <u>atomic masses</u>, like Newlands had, but he left gaps for the new elements that were yet to be found.
Mendeleev was also able to predict the properties of the missing elements. He was eventually proved right when these elements were discovered and were found to have the properties he had predicted. Also, Mendeleev did not stick too strictly to an order of increasing atomic mass.
When similar elements did not line up the order was swapped:

EXAMPLE

Tellurium atoms have a higher relative mass than iodine atoms, but the properties of both elements meant that tellurium was better placed in Group VI and iodine was better placed in Group VII. Although protons had not yet been discovered by lining up the elements in this way Mendeleev had actually put the atoms in order of increasing atomic number, or increasing number of protons.

QUICK TEST

1. How did Newlands arrange the elements?
2. Which element did Newlands find to have similar properties to lithium?
3 Why did Newlands' method develop problems?
4. How did Mendeleev's idea differ from Newlands'?
5. Why was tellurium placed before iodine?
6. How is the modern periodic table arranged?
7. What are the vertical columns in the periodic table called?
8. How many electrons are in the outer shell of all the members of Group I?
9. How many electrons are in the outer shell of all the members of Group III?
10. What are the horizontal rows in the periodic table called?

10. Periods
9. 3
8. 1
7. Groups
6. Increasing atomic number
5. Properties
4. He left gaps and made predictions
3. He left no gaps for elements that had not yet been discovered.
2. Sodium
1. In rows of seven.

TRANSITION METALS

Group	I	II													III	IV	V	VI	VII	0
Period																				
1								He 1												He 2
2	Li 3	Be 4													B 5	C 6	N 7	O 8	F 9	Ne 10
3	Na 11	Mg 12													Al 13	Si 14	P 15	S 16	Cl 17	Ar 18
4	K 19	Ca 20	Sc 21	Ti 22	V 23	Cr 24	Mn 25	Fe 26	Co 27	Ni 28	Cu 29	Zn 30			Ga 31	Ge 32	As 33	Se 34	Br 35	Kr 36
5	Rb 37	Sr 38	Y 39	Zr 40	Nb 41	Mo 42	Tc 43	Ru 44	Rh 45	Pd 46	Ag 47	Cd 48			In 49	Sn 50	Sb 51	Te 52	I 53	Xe 54
6	Cs 55	Ba 56	57 – 71*	Hf 72	Ta 73	W 74	Re 75	Os 76	Ir 77	Pt 78	Au 79	Hg 80			Tl 81	Pb 82	Bi 83	Po 84	At 85	Rn 86
7	Fr 87	Ra 88	89 – 103**																	

The transition metals are found in the <u>middle</u> <u>section</u> of the periodic table.
<u>Iron</u>, <u>nickel</u>, <u>copper</u> and <u>platinum</u> are examples.
All the transition metals have characteristic properties. They all have:
• high melting points
• high density.
• They are also strong, tough and hard-wearing and form coloured compounds (e.g. pottery glazes).

COPPER

• Copper is a <u>good</u> <u>conductor</u> of heat and electricity. It can be bent easily and does not <u>corrode</u>.
• Copper is used for electrical wiring because it can be bent into shape and is a good conductor of electricity.
• Copper is also used for making water pipes because it does not corrode and can be bent into shape without fracturing.

IRON

• Iron is <u>strong</u> but <u>brittle</u>.
• Iron is often made into <u>steel</u>.
• Steel is strong and cheap and is used in vast quantities; unfortunately it is also <u>heavy</u> and may <u>rust</u>.
• Iron and steel are useful <u>structural</u> materials. Bridges, buildings, ships, cars and trains are all constructed from these materials.
• Iron is a useful <u>catalyst</u> and is used in the Haber process.
• Stainless steel does not rust, but is more expensive to produce than iron.

NICKEL

• Nickel is <u>hard</u>, <u>shiny</u> and <u>dense</u>.
• It is used to make <u>coins</u>.
• Nickel is also used as a catalyst in the manufacture of margarine.

METAL STRUCTURE

- The <u>free</u> <u>electrons</u> hold the atoms together in a regular structure and give metals their special properties.
- The free electrons allow metals to <u>conduct</u> heat and electricity.
- The free electrons also allow the atoms to slide over each other, without breaking.
- This means that metals can be drawn into wires and hammered into shape.

Examiner's Top Tip
Remember the uses for the different metals featured here.

TRANSITION METALS

Metals have a <u>giant</u> structure. The electrons in the highest energy shells (outer electrons) are free to move through the whole structure.

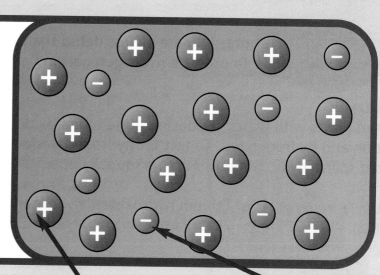

positive metal irons   sea of negative electrons

QUICK TEST

1. Why are metals able to conduct heat and electricity?
2. In which part of the periodic table are the transition metals found?
3. What are the characteristic properties of transition metals?
4. Why is copper used for electrical wiring?
5. Why is copper used for water pipes?
6. Why is iron often made into steel?
7. Give some uses of iron and steel?
8. In which process is iron used as a catalyst?
9. Which items are often made from nickel?
10. Nickel is used as a catalyst for the manufacture of which food stuff?

Examiner's Top Tip
Aluminium is not a transition metal, but it is another useful metal. It has a low density (it is light for its size) and can be used to make aeroplanes and drinks cans.

10. Margarine
9. Coins
8. Haber process
7. Bridges, buildings, ships, cars and trains
6. Iron is brittle.
5. Does not corrode or fracture
4. Good conductor, can be bent
3. High melting point , high density, shiny, hard-wearing and form coloured compounds, catalysts
2. Middle section
1. Free electrons

STRUCTURE

Down the group each element becomes less reactive.
The halogens have <u>coloured</u> <u>vapours</u>.
Down the group the colour of the vapour
gets <u>darker</u>.

- Melting and boiling points <u>increase</u> down the group
  (the first two are gases, the next – bromine – is a
  liquid, and iodine is a solid).
- All the halogens are <u>poisonous</u> and should only be
  used in a fume cupboard.
- Halogens react with metals to form <u>compounds</u>, in
  which the chloride, bromide or iodide ion carries
  a 1– charge.
- Halogens can also <u>react</u> <u>with</u> <u>non-metals</u> to form
  <u>covalently</u> <u>bonded</u> <u>compounds</u>.
- They are brittle and crumbly when solid.
- They are poor conductors of heat and electricity.

	V	VI	VII	0
				He 2
	O 8	F 9	Ne 10	
	S 16	Cl 17	Ar 18	
	Se 34	Br 35	Kr 36	
	Te 52	I 53	Xe 54	
	Po 84	At 85	Rn 86	

DISPLACEMENT REACTIONS

- A more reactive halogen will displace a less reactive halogen from its solution.
- Chlorine will displace bromine and iodine.
- Bromine will displace iodine, but not chlorine.

EXAMPLE
<u>Chlorine</u> will displace iodine from a solution of <u>potassium</u> <u>iodide</u>.
chlorine + potassium iodide ⟹ iodine + potassium chloride
$Cl_2(g)$ + $2KI(aq)$ ⟹ $I_2(aq)$ + $2KCl(aq)$

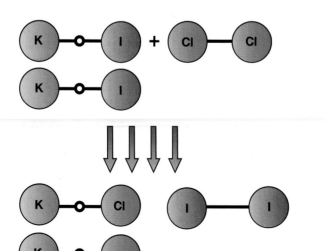

Examiner's Top Tip
All Group VII atoms
form diatomic
molecules.

Examiner's Top Tip
All atoms in Group
VII have seven
outer electrons.

THE HALOGEN FAMILY

FLUORINE
- *Fluorine is a <u>very</u> <u>poisonous</u>, pale yellow <u>gas</u>.*

CHLORINE
- *Chlorine is a <u>poisonous</u>, pale green <u>gas</u>.*
- *Chlorine is used in water purification and bleaching.*

BROMINE
- *Bromine is a <u>poisonous</u>, dense, brown <u>liquid</u>.*

IODINE
- *Iodine is a dark grey, crystalline <u>solid</u> or a purple <u>vapour</u>.*
- *Iodine solution is used as an antiseptic.*

Examiner's Top Tip
Learn the uses of the different halogens.

GROUP VII – THE HALOGENS

The halogens all have <u>seven</u> electrons in their outer shell

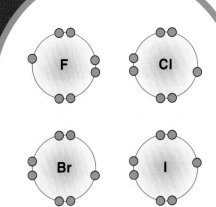

a model showing the outer shell of electrons

Examiner's Top Tip
Compare the trends for Group VII with the trends for Group I.

QUICK TEST

1. What is the name used for Group VII?

2. What is the trend in the size of atoms down Group VII?

3. What is the trend in reactivity down Group VII?

4. What safety precautions should be used for halogens?

5. What is the trend in the melting and boiling points down Group VII?

6. What is the state of the first four halogens?

7. What is the trend in the colour of the elements down Group VII?

8. What is chlorine used for?

9. What is iodine solution used for?

10. What would happen if chlorine gas was reacted with potassium bromide?

10. Displacement gives potassium chloride + bromine.
9. Antiseptic
8. Water purification/bleaching
7. Colour gets darker (pale yellow, green, brown, dark grey/purple)
6. Gas, gas, liquid, solid
5. Increase
4. Fume cupboard, etc
3. Decreases
2. Increases
1. Halogens

GROUP 0 –
THE NOBLE GASES

The noble gases are <u>unreactive</u>.
They are sometimes called 'inert' because they <u>do not react</u>.
This is because they <u>all</u> <u>have</u> <u>a</u> <u>full</u> <u>outer</u> <u>shell</u> <u>of</u> <u>electrons</u>.

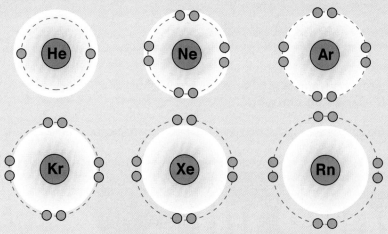

a model showing the outer shell of electrons of the noble gases

USES OF THE NOBLE GASES

HELIUM

- Helium is used in <u>balloons</u> and in <u>airships</u>, because it is <u>less dense</u> than air (and <u>not flammable</u> like hydrogen).

ARGON

- Argon is used in light bulbs (<u>filament lamps</u>).
- Surrounding the hot filament lamp with inert argon stops it from burning away.

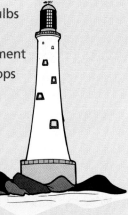

NEON

- Neon is used in electrical discharge tubes in advertising signs.

KRYPTON

- Krypton is used in <u>lasers</u>.

THE TRENDS IN THE NOBLE GASES, GROUP 0

As you go down the group:
• **density increases**
• **boiling point increases**
• **they are <u>colourless</u>, <u>monatomic</u> <u>gases</u>**
(they exist as individual atoms rather than as diatomic molecules as other gases do).

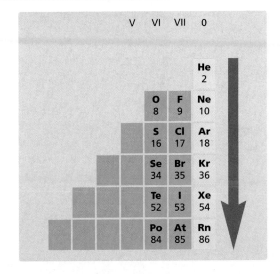

PHYSICAL PROPERTIES OF NOBLE GASES

• *All the members of Group 0 are <u>gases at room temperature</u>.*
• *If they are <u>cooled down</u> they can become <u>liquids</u> and eventually <u>solids</u>.*
• *In a solid state they are <u>brittle</u> and <u>crumbly</u> (typical for non-metals).*
• *They are also poor conductors of both electricity and heat.*

Examiner's Top Tip
Learn the trends down this non-metal group, and the uses of the noble gases.

QUICK TEST

Examiner's Top Tip
There is room for up to two electrons in the first shell, and up to eight in the second shell.

1. Why are the noble gases unreactive?
2. As you go down the group, what happens to density and boiling point?
3. Draw the (outer shell) electron structure of helium.
4. Draw the (outer shell) electron structure of argon.
5. What does 'monatomic' mean?
6. What is helium used for?
7. Why is it used?
8. What is neon used for?
9. What is argon used for?
10. What is krypton used for?

10. Lasers
9. Surrounds filaments in light bulbs
8. Electrical discharge tubes
7. It is less dense than air and not flammable.
6. Balloons and airships
5. Existing as individual atoms
4. See diagram opposite
3. See diagram opposite
2. They increase
1. They have a full outer shell of electrons.

SODIUM CHLORIDE

Sodium chloride (common salt) is an important resource.
Sodium chloride is a compound of a Group I metal (sodium) and a Group VII non-metal (chlorine).
Salt is found in large quantities in the <u>sea</u> and also in vast underground <u>deposits</u> laid down as ancient seas evaporated.
Rock salt is used on <u>icy</u> <u>roads</u>.
The salt <u>lowers</u> the freezing point of water from 0°C to about –5°C.
This means that the water on the roads does not freeze to form ice until the temperature is much lower.

Examiner's Top Tip
Now cover this page and try to write down the uses of sodium chloride.

ELECTROLYSIS OF BRINE

ELECTROLYSIS OF SODIUM CHLORIDE SOLUTION (BRINE)

Sodium chloride dissolved in water is called <u>brine</u>.
Electrolysis of concentrated brine solution is an important industrial process.

Examiner's Top Tip
Learn what happens when a concentrated sodium chloride solution is electrolysed.

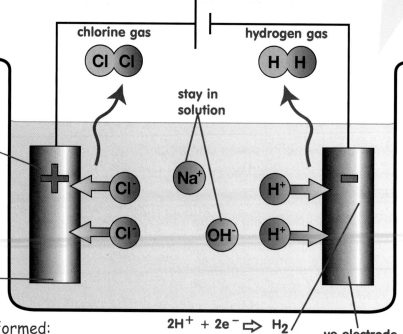

chlorine gas

hydrogen gas

stay in solution

$2Cl^- \Rightarrow Cl_2 + 2e^-$

+ve electrode

$2H^+ + 2e^- \Rightarrow H_2$

–ve electrode

Three important products are formed:
• <u>Chlorine</u> gas is released at the positive electrode.
• <u>Hydrogen</u> gas is formed at the negative electrode.
• A solution of <u>sodium</u> <u>hydroxide</u> is also produced.
Each of these products can be used to make other useful materials.

USEFUL PRODUCTS FROM THE ELECTROLYSIS OF BRINE

CHLORINE

This is used:

- **to make <u>bleach</u>**
- **to <u>sterilise</u> <u>water</u>**
- **to produce <u>hydrochloric</u> <u>acid</u>**
- **in the production of <u>PVC</u>**

HYDROGEN

This is used in the manufacture of <u>margarine</u>.

SODIUM HYDROXIDE

This is an alkali used in paper-making and the manufacture of many products including:

- **<u>soaps</u> and <u>detergents</u>**
- **<u>rayon</u> and <u>acetate</u> fibres**

QUICK TEST

1. What groups of the periodic table do sodium and chlorine belong to?

2. Where is salt found?

3. Why are roads 'salted'?

4. What is brine?

5. During the electrolysis of brine what is formed at the positive electrode?

6. What is formed at the negative electrode?

7. Which other useful chemical is made?

8. What are the uses of chlorine?

9. What are the uses of hydrogen?

10. What are the uses of sodium hydroxide?

Examiner's Top Tip
What are the uses of the products of the electrolysis of brine?

10. Soap, detergents, paper, rayon, acetate
9. It is used in the manufacture of margarine
8. Bleach, sterilise water, hydrochloric acid, PVC
7. Sodium hydroxide
6. Hydrogen
5. Chlorine
4. Sodium chloride dissolved in water.
3. Salt lowers the freezing point of water.
2. Sea, underground deposits
1. I and VII

EXAM QUESTIONS — Use the questions to test your progress. Check your answers on pages 94–95.

1. The three boxes show the particle arrangements in a solid, liquid and a gas.
   Which diagram represents which state of matter?

   a   b   c

   ....

   ....

   ....

2. What is the charge on a proton?

   ....

3. What is the chemical formula of sodium chloride?

   ....

4. What is the name given to the vertical columns in the periodic table?

   ....

5. Name three elements from Group I.

   ....

6. Name three elements from Group VII.

   ....

7. Name three elements from Group 0.

   ....

8. a) In which group of the periodic table is this element found?

   ....

   b) How many protons does an atom of this element have?

   ....

9. A neutral atom gains two electrons. What is the charge on the ion formed?

   ....

   ....

10. How are the elements arranged in the periodic table?

   ....

   ....

11. An element has seven electrons. What is its electron structure and to which group does it belong?

   ....

12. Sodium is in Group I of the periodic table. What does this tell you about the electronic structure of sodium?

   ....

   ....

13. fluorine – chlorine – bromine – iodine
These four elements belong to Group VII; which one is a liquid at room temperature?
....

14. Name the type of bond formed by the strong force of attraction between oppositely charged ions.
....

....

15. Why do all the elements in Group VII have similar chemical properties?
....

....

16. What are the three products made by the electrolysis of concentrated sodium chloride solution?
....

....

17. Describe the particles found in the nucleus.
....

....

....

18. Complete the table below:

	number of protons	number of electrons	number of neutrons	electron structure
$^{12}_{6}C$				
$^{14}_{6}C$				

19. Sodium chloride is an ionic compound.
    Draw a 'dot and cross' diagram to show how sodium chloride is formed from its elements.

20. a) Magnesium has an electronic structure of 2, 8, 2. To which group does it belong?
....

b) Magnesium can form ionic compounds. What charge do magnesium ions carry in such compounds?

....

How did you do?

1–5	correct ....start again
6–10	correct ....getting there
11–15	correct ....good work
16–20	correct ....excellent

COMMON TESTS AND SAFETY HAZARDS

COMMON TESTS

CARBON DIOXIDE

The gas is <u>bubbled</u> through <u>limewater</u>.
Carbon dioxide turns limewater <u>milky</u>.

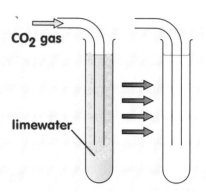

CO_2 gas

limewater

HYDROGEN

If a lighted splint is nearby hydrogen will burn
with a '<u>squeaky pop</u>'.

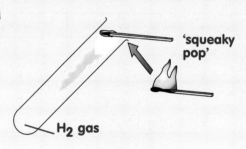

'squeaky pop'

H_2 gas

CHLORINE

Chlorine <u>bleaches</u> damp litmus paper.

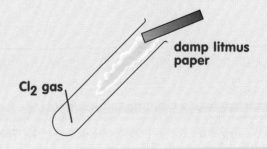

damp litmus paper

Cl_2 gas

OXYGEN

Oxygen <u>relights</u> a glowing <u>splint</u>.

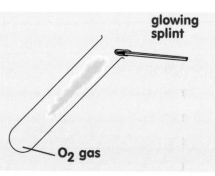

glowing splint

O_2 gas

SAFETY HAZARDS

OXIDISING
- Provides <u>oxygen</u> which allows <u>other</u> <u>materials</u> to burn <u>more</u> <u>fiercely</u>.

HIGHLY FLAMMABLE
- Catches fire easily.

TOXIC
- Can cause <u>death</u> if <u>swallowed</u>, <u>breathed</u> in or <u>absorbed</u> through the skin.

HARMFUL
- Similar to <u>toxic</u> but less dangerous.

CORROSIVE
- Attacks and destroys <u>living</u> <u>tissues</u>, including <u>eyes</u> and <u>skin</u>.

IRRITANT
- Not corrosive, but can cause reddening or blistering of the skin.

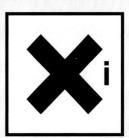

QUICK TEST

1. Sketch the hazard symbol for oxidising.
2. Sketch the hazard symbol for highly flammable.
3. Sketch the hazard symbol for toxic.
4. Sketch the hazard symbol for corrosive.
5. Sketch the hazard symbol for irritant.
6. Sketch the hazard symbol for harmful.
7. What is the test for carbon dioxide?
8. What is the test for hydrogen?
9. What is the test for chlorine?
10. What is the test for oxygen?

Examiner's Top Tip
None of this is hard, it is just a case of learning all the points.

10. Glowing splint relights.
9. Damp litmus paper bleached
8. Lighted splint gives 'squeaky pop'.
7. Gas is bubbled through limewater which turns milky.
6. See above
5. See above
4. See above
3. See above
2. See above
1. See above

TEMPERATURE

If the temperature is increased, particles move quicker.
Increasing the temperature increases the <u>rate</u> of <u>reaction</u> because the particles collide more often and with more energy so there are more successful collisions, and the rate increases.
a <u>low</u> <u>temperature</u> slows reaction rate down

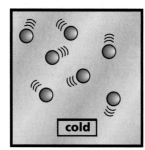

cold

a <u>high</u> <u>temperature</u> speeds reaction rate up

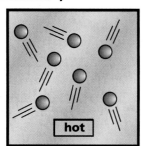

hot

INCREASING THE SURFACE AREA

The greater the <u>surface</u> <u>area</u>, the more chance of collisions occurring, so the faster the rate of reaction.

small surface area (larger pieces)

large surface area (smaller pieces)

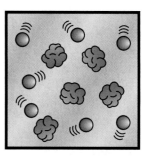

ADDING A CATALYST

- A catalyst <u>increases</u> the rate of reaction, but is not itself used up in the reaction.
- Catalysts are specific to certain reactions.

Examiner's Top Tip
When analysing graphs, remember that the reaction is over when the graph levels out.

ANALYSING RATES OF REACTION

The rate of a chemical reaction can be measured by:
- how fast the <u>products</u> are being made
- how fast the <u>reactants</u> are being used up.

The graph shows the amount of <u>product</u> made in three experiments.
The graph is <u>steepest</u> at the start of the reaction for all three experiments; it then starts to <u>level out</u> as the reactant particles get used up.
When the graph becomes level, the reaction has finished. The graph shows that <u>Experiment 2</u> is faster than <u>Experiment 1</u>.
This could have been due to Reaction 2 having a higher temperature, a greater concentration of reactant particles (or pressure for gases), a greater surface area or a catalyst being added to it.
Experiments 1 and 2 make the <u>same</u> <u>amount</u> of the product so they both had the <u>same</u> <u>amounts</u> of reactants at the start of the reaction.
In Experiment 3 only <u>half</u> as much product was made.
This shows that the amount of the reactants was <u>less</u> at the start of Reaction 3.

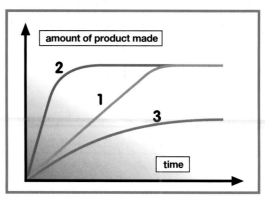

amount of product made

2

1

3

time

INCREASING THE CONCENTRATION OF DISSOLVED REACTANTS

The greater the concentration, the more reactant particles there are in the solution. There will be more collisions and so the reaction rate is increased. For gases, increasing the pressure has the same effect as increasing the concentration of dissolved particles.

low concentration of particles

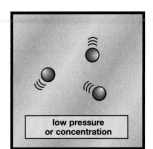

low pressure or concentration

high concentration of particles

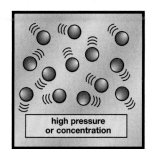

high pressure or concentration

RATES OF REACTION

Rates of reaction can be
slow or fast:

rusting ↙   ↘ explosions

Reaction Energy

A chemical reaction can only occur if the reacting particles collide with enough energy (activation energy) to react. If they collide and do not have enough energy they bounce apart and do not react.

QUICK TEST

1. What happens if two particles collide but they do not have enough activation energy?
2. What happens to the rate of a reaction if the temperature is decreased?
3. What happens to the rate of a reaction if the concentration of the dissolved reactants is increased?
4. What happens to the rate of a reaction if the pressure of gas reactants is increased?
5. What happens to the rate of a reaction if the surface area of a reactant is increased?
6. What does a catalyst do to the rate of reaction?
7. Why can catalysts be reused?
8. How can the rate of a chemical reaction be measured?
9. Give four ways in which you could increase the rate of a chemical reaction.
10. Give three ways in which you could decrease the rate of a chemical reaction.

10. Decrease temperature; decrease concentration (pressure); decrease surface area
9. Increase temperature; increase concentration (pressure); increase surface area; add a catalyst
8. How fast reactant used up/product is made
7. Not used up during the reaction
6. Increases
5. Increases
4. Increases
3. Increases
2. Decreases
1. They do not react (bounce apart)

Examiner's Top Tip
Remember: small pieces have a larger surface area.

CATALYSTS AND ENZYMES

Catalysts:
- **can increase the rate of a chemical reaction**
- **are not used up during a reaction, so they can be used again**
- **are important in industry, as speeding up reactions reduces costs.**

HOW THEY WORK

A catalyst can increase the rate of reaction, so the products are made faster.

Examiner's Top Tip
Catalysts are specific to a particular reaction, so only name a catalyst if you are sure that it works for that reaction.

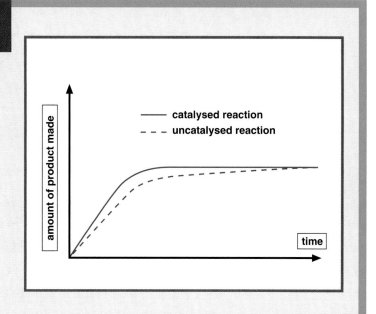

amount of product made

— catalysed reaction
- - - uncatalysed reaction

time

ENZYMES

Enzymes are biological catalysts; they are protein molecules. Enzymes work well in warm conditions. If the temperature is too low the enzymes stop working as quickly.

Freezing food reduces the enzymes' activity, so the food remains fresh for longer. However, the enzymes will not have been destroyed, and on warming the food will continue to deteriorate. If the temperature is too high the enzyme will be denatured (damaged).

Enzymes usually stop working at temperatures above about 45ºC. Different enzymes work best at different pHs.

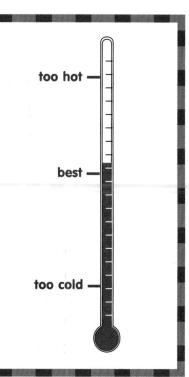

too hot —

best —

too cold —

USES OF CATALYSTS

Catalysts and enzymes are used in many everyday situations:

BREAD-MAKING (FERMENTATION)
- In **fermentation**, yeast converts **sugar** into **carbon dioxide**, **ethanol** and **energy**.
- During bread-making, yeast produces **bubbles** of carbon dioxide which gives the bread its **light** **texture**.

$$\text{glucose} \xrightarrow{\text{yeast}} \text{carbon dioxide} + \text{an alcohol} + \text{energy}$$

BREWING (FERMENTATION)
- Wine and beer are also made by **fermentation**.
- **Glucose** (sugar) from **fruit**, **vegetables** and **cereals** is converted into **alcohol** and **carbon dioxide**.

YOGHURT MAKING
- In yoghurt-making **bacteria** are added to milk.
- These bacteria convert **lactose**, the sugar in milk, into **lactic acid**.

BIOLOGICAL DETERGENTS
- Biological detergents contain **protein-digesting** and **fat-digesting** enzymes (**proteases** and **lipases**) so they can get clothes cleaner.

INDUSTRIAL USES
- Enzymes that break down proteins (**proteases**) are used in the production of some baby foods.
- Enzymes that break down **starch** into sugar (**carbohydrases**) are used in soft-centred chocolates.
- **Isomerase** converts glucose into **fructose**. Fructose is much sweeter and so can be used in smaller quantities in slimming foods.

QUICK TEST

1. Why can catalysts be used again?
2. Why are catalysts used in industry?
3. Copper sulphate can act as a catalyst. Would you recommend it to speed up any reaction?
4. What are enzymes?
5. What happens to enzymes in cold conditions?
6. What happens to enzymes in hot conditions?
7. Why does bread have a light texture?
8. In brewing what ingredients can provide the glucose (sugar) for fermentation to occur?
9. Which enzymes break down proteins?
10. Which enzymes break down fats?

Examiner's Top Tip
Learn the different industrial uses of enzymes.

1. They are not used up in the reaction.
2. They speed up reactions and reduce costs.
3. No, catalysts are specific to a particular reaction
4. Biological catalysts
5. They stop working as quickly.
6. They are denatured
7. Due to bubbles of carbon dioxide being produced during fermentation
8. Fruit/vegetables/cereals
9. Proteases
10. Lipases

EXOTHERMIC REACTIONS

In **exothermic** reactions energy, usually in the form of heat, is **given out** to the surroundings. This normally causes a **rise** in temperature.
Burning fuel is an **exothermic reaction** which gives out a lot of heat.

ENDOTHERMIC REACTIONS

In <u>endothermic</u> reactions energy, normally in the form of heat, is <u>taken in</u> from the surroundings. This is shown by a <u>decrease</u> in temperature.

SIMPLE REVERSIBLE REACTIONS

The products can **react** to form the **original reactants**.
If the forward reaction is **exothermic** (gives out heat) the **backward** reaction is **endothermic** (takes heat in).
The amount of heat **given out** and **taken in** must be the **same**.

EXAMPLE
The **thermal decomposition** of hydrated copper sulphate (hydrated means 'with water', anhydrous means 'without water')

water vapour

In the **forward reaction** heat is taken in (**endothermic**).
*   hydrated copper sulphate ⇨ anhydrous copper sulphate + water
        (blue)   (white)

forward reaction

In the **backward reaction** heat is given out (**exothermic**).
*   anhydrous copper sulphate + water ⇨ hydrated copper sulphate
        (white)   (blue)

backward reaction

DYNAMIC EQUILIBRIUM

If a <u>reversible</u> <u>reaction</u> takes place in a <u>closed</u> <u>system</u> (where nothing can escape) eventually an <u>equilibrium</u> will be reached.

It is a <u>dynamic equilibrium</u>: both the <u>forward</u> and <u>backward</u> reactions are taking place at exactly the <u>same</u> <u>rate</u>.

The <u>conditions</u> will affect the <u>position</u> <u>of</u> <u>equilibrium</u> (that is, how much reactant and product are present at equilibrium).

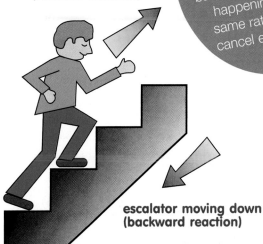

student trying to run up (forward reaction)

escalator moving down (backward reaction)

REVERSIBLE REACTIONS

Some chemical reactions are <u>reversible</u>; they can proceed in <u>both</u> <u>directions</u> (forwards and backwards).

- $A + B \rightleftharpoons C + D$

QUICK TEST

1. What is special about a reversible reaction?

2. If the forward reaction gives out energy, what type of reaction is it?

3. In the reverse reaction energy is taken in; what type of reaction is this?

4. What can be said about the amount of energy in each case?

5. Give the equation for a reaction that is endothermic.

6. Give the equation of a reaction that is exothermic.

6. Anhydrous copper sulphate + water —> hydrated copper sulphate
5. Hydrated copper sulphate —> anhydrous copper sulphate + water
4. It is the same.
3. Endothermic
2. Exothermic
1. It can proceed in either direction.

THE HABER PROCESS

THE HABER PROCESS

This is an example of a <u>reversible reaction</u>.

- nitrogen and hydrogen ⇌ ammonia

Some of the nitrogen and hydrogen react to form ammonia. At the same time, some of the ammonia breaks down into nitrogen and hydrogen.

Not all of the hydrogen and nitrogen are converted to ammonia, giving a mixture of hydrogen, nitrogen and ammonia. When this mixture is <u>cooled</u> the ammonia <u>liquefies</u> and is removed. The remaining nitrogen and hydrogen is recycled to reduce costs.

INDUSTRIAL CONDITIONS
- high pressure (200 atmospheres)
- quite high temperature (450°C)
- an iron catalyst.

The catalyst speeds up the rate of reaction and so reduces the cost of producing the ammonia. These conditions produce a reasonable amount of ammonia fairly quickly.

Nitrogen and hydrogen are mixed

Iron catalyst Temperature of 450°C Pressure of 200 atmospheres

Unreacted hydrogen and nitrogen are recycled

Ammonia is produced

USES OF AMMONIA

Ammonia can be oxidised to produce <u>nitric acid</u>.
Ammonia gas reacts with oxygen in the air over a <u>hot platinum catalyst</u>.

- ammonia + oxygen ⇨ nitrogen monoxide + water

The nitrogen oxide is <u>cooled</u>, and then reacted with <u>water</u> and more <u>oxygen</u> to form <u>nitric acid</u>.

- nitrogen monoxide + oxygen + water ⇨ nitric acid

The nitric acid can be neutralised with <u>ammonia</u> to make <u>ammonium nitrate</u>.
Ammonia can also be reacted with <u>sulphuric acid</u> to make <u>ammonium sulphate</u>.
These are both popular <u>fertilisers</u>.

- **Ammonia** is produced by the **Haber** **process**.
- Ammonia is made of **nitrogen** and **hydrogen**.
- The hydrogen is obtained from **natural gas** and the nitrogen is obtained from the **air**.

INTERPRETING CHEMICAL EQUATIONS

nitrogen + hydrogen ⇨ ammonia

$$N_2(g) + 3H_2(g) \Rightarrow 2NH_3(g)$$

nitrogen   **hydrogen**   **ammonia**

NH_3 means 1 nitrogen and 3 hydrogen atoms.

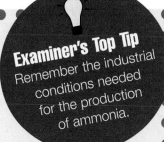

QUICK TEST

1. What does the Haber process produce?
2. From where is the hydrogen obtained?
3. From where is the nitrogen obtained?
4. Why is this described as a reversible reaction?
5. How is the ammonia removed?
6. Which catalyst is used in this process?
7. At what pressure is the process carried out?
8. At what temperature is the process carried out?
9. In what proportion should the hydrogen and nitrogen be mixed?

9. 3:1
8. 450°C
7. 200 atmospheres
6. Iron
5. On cooling it liquefies.
4. Nitrogen reacts with hydrogen to make ammonia; ammonia breaks up to form nitrogen and hydrogen.
3. Air
2. Natural gas
1. Ammonia

RELATIVE FORMULA MASS

RELATIVE ATOMIC MASS (RAM)

The relative atomic mass (RAM) is used to compare the masses of different atoms. The relative atomic mass of an element is the average mass of its isotopes compared with an atom of $^{12}_{6}C$.

4 ← mass number
2 ← atomic number (or proton number)

$^{4}_{2}He$

RAM of helium = 4

$^{24}_{12}Mg$

RAM of magnesium = 24

RELATIVE FORMULA MASS

The relative formula mass of any molecule is worked out by adding together the relative atomic masses of all the atoms in the molecule.

* For carbon dioxide, CO_2:

$$C \; O_2$$
$$12 + (2 \times 16) = 44$$

The relative formula mass of CO_2 is 44.

* For water, H_2O:

$$H_2 \; O$$
$$(2 \times 1) + 16 = 18$$

The relative formula mass of H_2O is 18.

* For ammonia, NH_3:

$$N \; H_3$$
$$14 + (3 \times 1) = 17$$

The relative formula mass of NH_3 is 17.

CALCULATING THE PERCENTAGE COMPOSITION OF AN ELEMENT IN A COMPOUND

Percentage mass of an element in a compound $= \dfrac{\text{relative atomic mass x no. of atoms}}{\text{relative formula mass}} \times 100\%$

EXAMPLE

Ammonium nitrate, NH_4NO_3, is used as a fertiliser.
Find the <u>percentage</u> <u>composition</u> of nitrogen in this compound.

- RAM of N = 14
- RAM of H = 1
- RAM of O = 16

The formula mass of NH_4NO_3 is:
- $14 + (4 \times 1) + 14 + (3 \times 16) = 80$

Percentage of nitrogen $= \dfrac{14 \times 2}{80} \times 100\% = 35\%$

- <u>Ammonium</u> <u>nitrate</u> <u>is</u> <u>35%</u> <u>nitrogen</u>

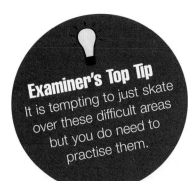

Examiner's Top Tip
It is tempting to just skate over these difficult areas but you do need to practise them.

QUICK TEST

1. Find the RAM of carbon, C.
2. Find the relative formula mass of nitrogen molecules, N_2.
3. Find the relative formula mass of oxygen molecules, O_2.
4. Calculate the relative formula mass of carbon monoxide, CO.
5. Calculate the relative formula mass of copper sulphate, $CuSO_4$.
6. Calculate the relative formula mass of calcium carbonate, $CaCO_3$.
7. Calculate the percentage of hydrogen in ammonium nitrate, NH_4NO_3.
8. Calculate the percentage of oxygen in ammonium nitrate, NH_4NO_3.
9. Calculate the percentage of hydrogen in water, H_2O.
10. Calculate the percentage of sulphur in sulphur dioxide, SO_2.

10. 50%
9. 11%
8. 60%
7. 5%
6. 100
5. 159.5
4. 28
3. 32
2. 28
1. 12

BALANCING THE EQUATION

When <u>hydrogen</u> burns in <u>oxygen</u>, <u>water</u> is made.

- <u>Hydrogen</u> + <u>oxygen</u> ⟹ <u>water</u>

- H_2 + O_2 ⟹ H_2O

The <u>formulae</u> are correct, but the equation is <u>not</u> balanced because there are different numbers of atoms on each side of the equation. The formulae <u>cannot</u> be changed, but the numbers in front of the formulae <u>can</u> be changed.

HOW TO BALANCE AN EQUATION

Looking at the equation we can see that there are <u>two</u> oxygen atoms on the left-hand side but only <u>one</u> on the right-hand side.

So a <u>2</u> is placed <u>in</u> <u>front</u> of the H_2O:

- $H_2 + O_2$ ⟹ $2H_2O$

Now the oxygen atoms are balanced, but while there are <u>two</u> hydrogen atoms on the left-hand side there are now <u>four</u> hydrogen atoms on the right-hand side.

So a <u>2</u> is placed in front of the H_2:

- $2H_2 + O_2$ ⟹ $2H_2O$

- *The equation is then balanced.*

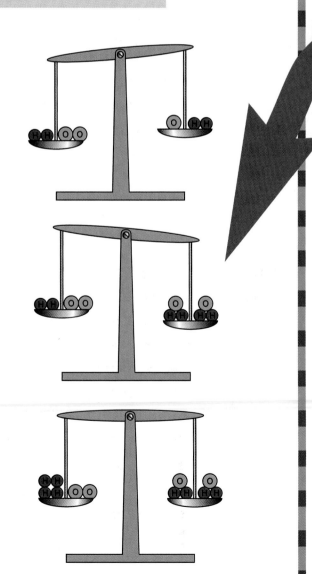

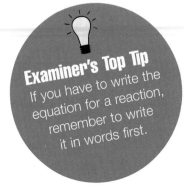

Examiner's Top Tip
If you have to write the equation for a reaction, remember to write it in words first.

STATE SYMBOLS

State symbols can be added to an equation to give extra information.
They show what state the reactants and products are in.
The symbols are:

- (s) for solid
- (l) for liquid
- (g) for gas
- (aq) for aqueous, or dissolved in water

EXAMPLE

magnesium + oxygen   magnesium oxide

- $2Mg(s)$ + $O_2(g)$ $2MgO(s)$

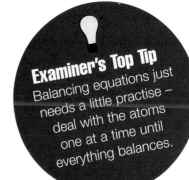

Examiner's Top Tip
Balancing equations just needs a little practise – deal with the atoms one at a time until everything balances.

BALANCING EQUATIONS

- **Symbol equations show the number of atoms.**
- **There must be the same number of atoms on both sides of the equation: atoms cannot be created or destroyed.**

magnesium + oxygen		magnesium oxide
2Mg + **O₂**		**2MgO**
●● + ∞		●○ ●○

Examiner's Top Tip
When balancing an equation always check that the formulae you have written down are correct.

QUICK TEST

1. How many calcium atoms are present in $CaCO_3$?
2. How many carbon atoms are present in $CaCO_3$?
3. How many oxygen atoms are present in $CaCO_3$?
4. Why must there be the same number of atoms on both sides of the equation?
5. Balance the equation $Na(s)$ + $Cl_2(g)$ ⇨ $NaCl(s)$.
6. Balance the equation $H_2(g)$ + $Cl_2(g)$ ⇨ $HCl(g)$.
7. Balance the equation $C(s)$ + $CO_2(g)$ ⇨ $CO(g)$.
8. What does the state symbol (l) indicate?
9. What does the state symbol (aq) indicate?
10. Add the state symbols to this equation for the
    thermal decomposition of calcium carbonate:
    $CaCO_3$ ⇨ CaO + CO_2.

10. $CaCO_3(s)$ ⇨ $CaO(s)$ + $CO_2(g)$
9. Aqueous
8. Liquid
7. $C(s)$ + $CO_2(g)$ ⇨ $2CO(g)$
6. $H_2(g)$ + $Cl_2(g)$ ⇨ $2HCl(g)$
5. $2Na(s)$ + $Cl_2(g)$ ⇨ $2NaCl(s)$
4. Atoms cannot be created or destroyed.
3. 3
2. 1
1. 1

THERMAL DECOMPOSITION

In **thermal** **decomposition** reactions a **substance** is broken down into **simpler** **substances** by heating.

EXAMPLE
The thermal decomposition of **limestone**:

- calcium carbonate $\overset{\text{HEAT}}{\Longrightarrow}$ calcium oxide + carbon dioxide

- $CaCO_3(s)$ $\Longrightarrow$ $CaO(s)$ + $CO_2(g)$

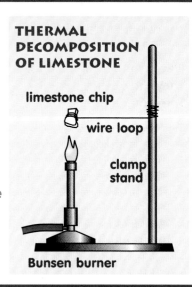

THERMAL DECOMPOSITION OF LIMESTONE

limestone chip

wire loop

clamp stand

Bunsen burner

TYPES OF REACTION

NEUTRALISATION

In a neutralisation reaction, an acid reacts with an alkali to form a salt and water.
- acid + alkali $\Longrightarrow$ salt + water
H^+ ions react with OH^- ions to form water.
- $H^+(aq)$ + $OH^-(aq)$ $\Longrightarrow$ $H_2O(l)$

EXAMPLE
The neutralisation of sodium hydroxide with ethanoic acid:
- sodium hydroxide + ethanoic acid $\Longrightarrow$ sodium ethanoate + water
- $NaOH(aq)$ + $CH_3COOH(aq)$ $\Longrightarrow$ $NaCH_3COO(aq)$ + $H_2O(l)$

bee stings are acidic and can be treated with sodium bicarbonate

OXIDATION

Many oxidation reactions involve the addition of oxygen.

EXAMPLE
The combustion of hydrogen:
- hydrogen + oxygen $\Longrightarrow$ water
- $2H_2(g)$ + $O_2(g)$ $\Longrightarrow$ $2H_2O(l)$

Oxidation also occurs during the electrolysis of aluminium oxide.

At the positive electrode:
- $2O^{2-}$ $\Longrightarrow$ O_2 + 4 electrons
Oxidation is loss of electrons.

hydrogen can be used as rocket fuel

REDUCTION

Reduction is the opposite of oxidation. Many reduction reactions involve the loss of oxygen. Metals are extracted from their oxides by reduction.

EXAMPLE
Iron oxide is reduced to iron in the blast furnace:
- iron + carbon $\Longrightarrow$ iron + carbon dioxide
  oxide   monoxide
- $Fe_2O_3(s)$ + $3CO(g)$ $\Longrightarrow$ $2Fe(l)$ + $3CO_2(g)$

Reduction also occurs during the electrolysis of aluminium oxide:
- Al^{3+} + 3 electrons $\Longrightarrow$ Al
Reduction is the gain of electrons.

iron is used to make cars and trains

REVERSIBLE REACTIONS

Reversible reactions can go in <u>both</u> directions (forwards and backwards).
In these reactions the products can <u>react</u> to produce the original reactants.

EXAMPLE

- ammonium chloride $\rightleftharpoons$ ammonia + hydrogen chloride

- $NH_4Cl(s)$ $\rightleftharpoons$ $NH_3(g)$ + $HCl(g)$

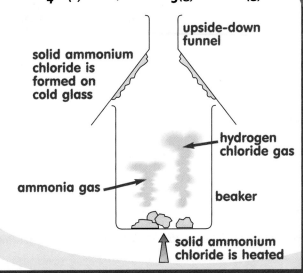

solid ammonium chloride is formed on cold glass

upside-down funnel

hydrogen chloride gas

ammonia gas

beaker

solid ammonium chloride is heated

EXOTHERMIC REACTIONS

Exothermic reactions <u>give out</u> energy, normally in the form of <u>heat</u>: They get <u>hotter</u>.

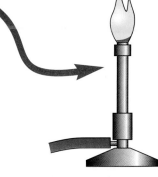

Examiner's Top Tip
Be familiar with all these different types of reaction.

EXAMPLE

The burning of fuels, such as methane, gives out a lot of <u>heat</u> energy:

- methane + oxygen $\Rightarrow$ carbon dioxide + water
- $CH_4(g)$ + $2O_2(g)$ $\Rightarrow$ $CO_2(g)$ + $2H_2O(l)$

Examiner's Top Tip
The combustion of hydrogen is an oxidation reaction; it is also an exothermic reaction. A reaction can be of more than one type.

ENDOTHERMIC REACTIONS

Endothermic reactions <u>take in</u> heat energy, normally in the form of <u>heat</u>. They get <u>cold</u>.

QUICK TEST

1. What happens during a thermal decomposition reaction?

2. What is produced when calcium carbonate is heated?

3. What happens during a neutralisation reaction?

4. Write the word equation for the neutralisation reaction between sodium hydroxide and ethanoic acid?

5. Most oxidation reactions involve the addition of which element?

6. Give an example of an oxidation reaction that takes place during electrolysis.

7. What type of reaction produces iron from iron oxide?

8. What is special about a reversible reaction?

9. Why is the burning of fuel an exothermic reaction?

10. What is special about endothermic reactions?

10. Takes in energy (heat); gets colder
9. Gives out energy (heat); gets hotter
8. Can go in both directions
7. Reduction
6. $2O^{2-} \longrightarrow O_2 + 4e^-$
5. Oxygen
4. Sodium hydroxide + ethanoic acid $\longrightarrow$ sodium ethanoate + water
3. Acid + alkali $\longrightarrow$ salt + water
2. Calcium oxide + carbon dioxide
1. Breakdown to simpler substances on heating

87

SEPARATION TECHNIQUES

- **In a mixture the constituent parts are <u>not</u> <u>joined</u> <u>together</u>.**
- **Mixtures can be <u>separated</u> quite easily.**

FILTRATION

- *Filtration is used to **<u>separate</u>** a mixture of a solid and a liquid.*
- *The mixture is poured through a **<u>filter</u>** **<u>paper</u>**. Only the liquid passes through and is called the **<u>filtrate</u>**. The solid is collected on the filter paper and is called the **<u>residue</u>**.*

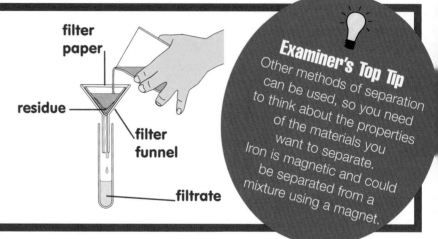

filter paper

residue

filter funnel

filtrate

FILTRATION AND EVAPORATION

- A mixture of <u>salt</u> and <u>sand</u> can be <u>separated</u> using these two techniques.

- When water is added and the mixture is stirred, the <u>soluble</u> <u>salt</u> will <u>dissolve</u>. The <u>insoluble</u> <u>sand</u> does not dissolve. The mixture can then be <u>filtered</u>: the <u>dissolved</u> <u>salt</u> passes through the filter, while the sand can be <u>collected</u> from the filter paper.

- Solutions of a <u>solvent</u> and a <u>solute</u> can be separated by <u>evaporation</u>. The <u>solvent</u>, in this case water, can be evaporated, leaving the solute (salt) behind. The salt forms <u>crystals</u>; this process is called <u>crystallisation</u>.

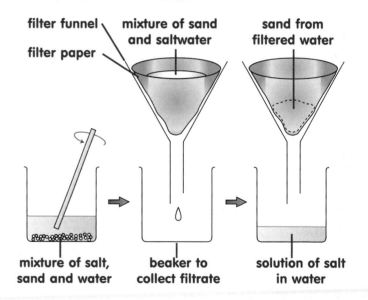

filter funnel

filter paper

mixture of sand and saltwater

sand from filtered water

mixture of salt, sand and water

beaker to collect filtrate

solution of salt in water

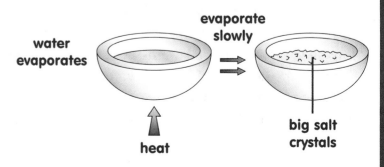

water evaporates

evaporate slowly

heat

big salt crystals

CHROMATOGRAPHY

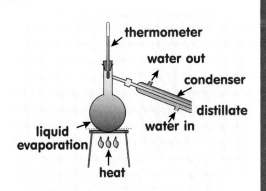

drop of ink
different dyes in ink
wick
water (solvent)

- <u>Chromatography</u> can be used to separate mixtures of <u>different</u> <u>coloured</u> <u>dyes</u>.
- The different dyes have <u>different</u> <u>solubilities</u>.
- This method can be used to find which dyes make up black ink.
- A spot of ink is placed on a piece of <u>filter</u> <u>paper</u> and this is placed in a beaker containing a small amount of <u>solvent</u>.
- The solvent travels across the filter paper, carrying the dyes with it. Each dye has a slightly different <u>solubility</u>, so travels a slightly different distance across the paper.
- This black dye contains red, blue and yellow dyes.

DISTILLATION

- Distillation can be used to separate a <u>solvent</u> from a <u>solution</u>.
- It can be used to separate water from a solution of salt and water.
  The solution is <u>heated</u>. The water boils and <u>water</u> <u>vapour</u> is formed. The water vapour cools and condenses to form <u>liquid</u> <u>water</u> which is collected in a beaker. This water is called '<u>distilled</u> <u>water</u>' and is very <u>pure</u>.

thermometer
water out
condenser
distillate
water in
liquid evaporation
heat

FRACTIONAL DISTILLATION

Fractional distillation can be used to separate a <u>mixture</u> of two or more liquids.
- It can be used to separate <u>alcohol</u> and <u>water</u>.
- The liquids still boil at their own boiling temperatures, even though they are now in a mixture. The alcohol boils at 78°C. Some water will also <u>evaporate</u>, but it will <u>condense</u> in the fractionating column and fall back into the flask.
- Only the alcohol passes into the condenser, where it forms <u>pure</u> <u>liquid</u> <u>alcohol</u> which is collected in the beaker.

QUICK TEST

1. In a mixture, are the constituent parts joined?
2. In a compound, are the constituent parts joined?
3. How should a mixture of solid and liquid be separated?
4. How can crystals of a salt be obtained from a mixture of salt and water?
5. Which technique should be used to separate a mixture of different coloured dyes?
6. Why do different dyes travel different distances?
7. How can water be separated from a mixture of salt and water?
8. What is the name of the piece of equipment in which the water vapour is turned back into liquid water?
9. How can a mixture of alcohol and water be separated?
10. In a mixture of alcohol and water, which boils first?

10. Alcohol
9. Through fractional distillation
8. Condenser
7. By distillation
6. They have different solubilities.
5. Chromatography
4. Through crystallisation, which will evaporate the water.
3. Filtration
2. Yes
1. No

89

EXAM QUESTIONS – Use the questions to test your progress. Check your answers on page 95.

1. What is the chemical formula of ammonia?

....

2. Is the burning of fossil fuels an exothermic or an endothermic reaction?

....

3. Ammonium nitrate is a fertiliser. It is produced by a neutralisation reaction between ammonia and which other chemical?

....

4. Why is a catalyst used in a chemical reaction?

....

5. Name the enzyme which breaks down starch to sugar.

....

6. Name the enzyme which breaks down glucose to fructose.

....

7. What do we call reactions that can proceed in either direction?

....

8. What is the name given to catalysts made by living things?

....

9. In a reaction magnesium is reacted with hydrochloric acid. What is the effect of increasing the surface area of the magnesium on the rate of the reaction?

....

10. When a gas is bubbled through limewater, the limewater turns milky. What is the name of the gas?

....

11. At what temperature would an enzyme work best?
    Choose one answer:

    −10ºC   10ºC   40ºC   70ºC

....

12. In an experiment, calcium carbonate reacts with hydrochloric acid. The graph shows how much carbon dioxide is produced.
a) When is the reaction over?

....

b) Draw on the graph what you would expect to see if the acid was heated.

....

13. A chemical bottle bears this label. What does it mean?

....

14. What happens to an enzyme if it is heated above about 45ºC?

....

....

15. Why does increasing the temperature increase the rate of a reaction?

....

....

16. Name the enzymes used in biological detergents.

....

17. What is special about a dynamic equilibrium?

....

18. Name the conditions used in the Haber Process.

....

19. What does the formula $CuSO_4$ (aq) represent?

....

20. Give the word equations for the conversion of ammonia into nitric acid.

....

....

....

....

How did you do?

1–5	correct	.start again
6–10	correct	.getting there
11–15	correct	.good work
16–20	correct	.excellent

EXAM QUESTIONS – Use the questions to test your progress. Check your answers on page 95.

1. Name the type of reaction which takes place when iron is extracted from iron oxide.

....

2. Name the type of reaction in which sodium hydroxide reacts with hydrochloric acid to produce sodium chloride and water.

....

3. Complete the word equation: calcium + oxygen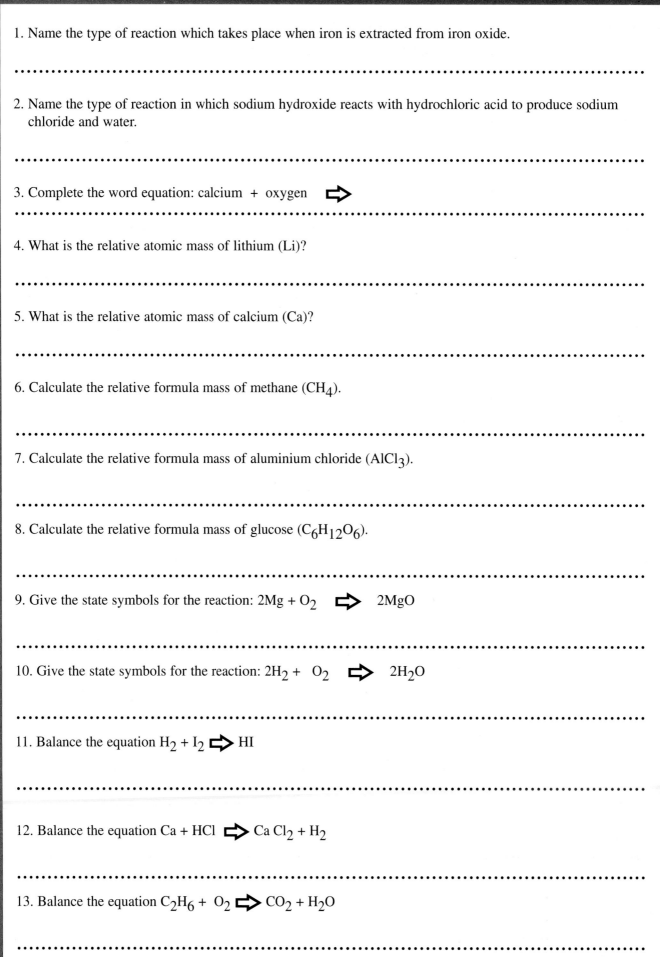
....

4. What is the relative atomic mass of lithium (Li)?

....

5. What is the relative atomic mass of calcium (Ca)?

....

6. Calculate the relative formula mass of methane (CH_4).

....

7. Calculate the relative formula mass of aluminium chloride ($AlCl_3$).

....

8. Calculate the relative formula mass of glucose ($C_6H_{12}O_6$).

....

9. Give the state symbols for the reaction: $2Mg + O_2$ ⟹ $2MgO$

....

10. Give the state symbols for the reaction: $2H_2 + O_2$ ⟹ $2H_2O$

....

11. Balance the equation $H_2 + I_2$ ⟹ HI

....

12. Balance the equation $Ca + HCl$ ⟹ $CaCl_2 + H_2$

....

13. Balance the equation $C_2H_6 + O_2$ ⟹ $CO_2 + H_2O$

....

14. Calcium carbonate has the formula $CaCO_3$. What atoms does it contain?

....

15. Copper sulphate has the formula $CuSO_4$. What atoms does it contain?

....

16. An atom loses an electron. What is the charge on the ion that is formed?

....

17. Water is a covalent compound. Draw a dot and cross diagram to represent it.

....

18. Explain what an isotope is.

....

19. Consider $^{24}_{12}Mg$, how many protons, neutrons and electrons are present in one atom of this element?

....

20. Why does hydrogen form H_2 molecules, while helium does not form molecules?

hydrogen   helium

....

....

21. Did intrusive, igneous or extrusive, igneous rocks cool more quickly?

....

22. Which gas found in polluted areas can cause acid rain?

....

23. Which salt is made when magnesium reacts with hydrochloric acid?

....

24. Which salt is made when zinc reacts with sulphuric acid?

....

25. What is the chemical name for slaked lime?

....

How did you do?

1–6	correct	....start again
7–13	correct	....getting there
14–19	correct	....good work
20–25	correct	....excellent

ANSWERS

Earth Materials

1. Igneous; sedimentary; metamorphic

2. Igneous

3. CFCs (chlorofluorocarbons).

4. Nitrogen

5. Sulphur dioxide

6. As carbonates and fossil fuels in sedimentary rocks

7. Metamorphic

8. a = crust; b = mantle; c = lithosphere; d = outer core; e = inner core

9. Calcium carbonate, $CaCO_3$

10. In making plastic bags and bottles

11. Plate tectonics

12. Convection currents caused by natural radioactive decay

13. Damage to buildings/trees/statues/plants and animals in lakes

14. a = youngest; d = oldest probably

15. Runny, easy to ignite; low boiling point

16. a) kerosene; b) petrol

17. Poly(ethene)

18. Poly(propene)

19. Carbon monoxide

20. Jigsaw fit; similar rock sequences; same fossil record

Metals

1. Magnesium, zinc, iron, copper

2. Hydrochloric acid/sulphuric acid/nitric acid

3. Sodium hydroxide/potassium hydroxide/ calcium hydroxide/ammonia solution

4. Universal indicator

5. The litmus paper would turn red.

6. Less reactive

7. Red

8. Sodium, zinc, iron, gold

9. Iron is more reactive and displaces the copper from the copper sulphate

10. Coke

11. Bauxite

12. Neutralisation

13. Electrolysis/reduction

14. So they can move

15. Carbon monoxide

16. a) Calcium chloride
    b) Potassium sulphate
    c) Sodium nitrate

17. Hydrogen and sodium hydroxide

18.

metal	solution		
	magnesium sulphate	copper sulphate	iron sulphate
magnesium	–	✓	✓
copper	X	–	X
iron	X	✓	–

19. Electrolysis

20. a) Alkaline   b) Acidic

Structure and Bonding

1. a) Solid   b) Gas   c) Liquid

2. 1+

3. NaCl

4. Groups

5. Li/Na/K/Rb/Cs/Fr

6. F/Cl/Br/I/At

7. He/Ne/Ar/Kr/Xe/Rn

8. a) VI   b) 16

9. 2-

10. In order of increasing atomic number

11. 2, 5; Group V

12. It has one outer electron

13. Bromine

14. Ionic bond

15. They have the same outer electron structure

16. Hydrogen, chlorine, sodium hydroxide

17. Neutrons: mass = 1, no charge; protons: mass = 1, charge 1+

18.	number of protons	number of electrons	number of neutrons	electron structure
$^{12}_{6}$C	6	6	6	2, 4
$^{14}_{6}$C	6	6	8	2, 4

19.

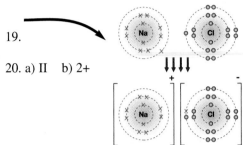

20. a) II   b) 2+

Chemical Change

1. NH_3

2. Exothermic

3. Nitric acid

4. It increases the rate of reaction (without being used up itself).

5. Carbohydrase

6. Isomerase

7. Reversible

8. Enzymes

9. It increases the rate of reaction

10. Carbon dioxide

11. 40°C

12. a) 4 min
    b) Line must be steeper, but not go higher than 0.09 g

13. Corrosive – will attack living tissue, including eyes and skin.

14. It is denatured.

15. The particles move faster; rate of collisions increases; more particles have enough energy to react; more successful collisions

16. Proteases and lipases

17. Both forward and backward reactions happen at the same rate.

18. 200 atmospheres, 450°C, iron catalyst

19. Aqueous copper sulphate or copper sulphate dissolved in water

20. Ammonia + oxygen ⇨ nitrogen monoxide+water
    nitrogen monoxide + oxygen + water ⇨ nitric acid

Mixed Questions

1. Reduction

2. Neutralisation

3. Calcium oxide

4. 7

5. 40

6. 16

7. 133.5

8. 180

9. $2Mg(s) + O_2(g)$ ⇨ $2MgO(s)$

10. $2H_2(g) + O_2(g)$ ⇨ $2H_2O(l)$

11. $H_2 + I_2$ ⇨ $2HI$

12. $Ca + 2HCl$ ⇨ $CaCl_2 + H_2$

13. $2C_2H_6 + 7O_2$ ⇨ $4CO_2 + 6H_2O$

14. 1 Ca 1 C and 3 O

15. 1 Cu 1 S and 4 O

16. 1+

17.

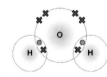

18. Isotopes are atoms of the same element that have the same number of protons (atomic number) but a different number of neutrons (mass number).

19. 12 protons 12 neutrons and 12 electrons

20. By covalent bonding two hydrogen atoms can gain a full, stable shell of electrons. Helium atoms already have a full, stable electron shell so do not react to form molecules.

21. Extrusive

22. Sulphur dioxide

23. Magnesium chloride

24. Zinc sulphate

25. Calcium hydroxide